D1535576

Voices of the Wild

Voices of the Wild

Animal Songs,

Human Din,

and the

Call to Save

Natural

Soundscapes

Bernie Krause

Yale UNIVERSITY PRESS/NEW HAVEN & LONDON

Published with assistance from the foundation established in memory of Amasa Stone Mather of the Class of 1907, Yale College.

Yale University Press books may be purchased in quantity for educational, business, or promotional use. For information, please e-mail sales. press@yale.edu (U.S. office) or sales@yaleup.co.uk (U.K. office).

Designed by Mary Valencia.
Set in Filosofia type by Integrated Publishing Solutions, Grand Rapids, Michigan.
Printed in the United States of America.

Library of Congress Cataloging-in-Publication Data
Krause, Bernard L.
Voices of the wild : animal songs, human din, and the call to save natural soundscapes / Bernie Krause.
      pages   cm — (The future series)
Includes bibliographical references and index.
ISBN 978-0-300-20631-9 (alk. paper)
    1. Acoustic phenomena in nature.   2. Nature sounds.   3. Bio-acoustics.   4. Sound—Physiological effect.   5. Sound—Psychological aspects.   6. Soundscapes (Music)   I. Title.
    QC809.A25K73 2015
    591.59'4—dc23   2014045425

A catalogue record for this book is available from the British Library.

This paper meets the requirements of ANSI/NISO Z39.48-1992 (Permanence of Paper).

10 9 8 7 6 5 4 3 2 1

Contents

Voices of the Wild

Introduction

In mid-January 2014, I spoke at a seminar in Yosemite
National Park. During a session break one sunny
afternoon with temperatures in the high sixties, during
what had become the driest and warmest continuous
period in four hundred years of California climate history, about a dozen participants walked the valley floor
of the park to Mirror Lake. The National Park Service
naturalist and guide who led us along the path was full
of useful information, enthusiastically tendered, as we
paused numerous times along the route to hear his ac-
counts of the geological features and characteristic life

forms that surrounded us. The effortless hike of a mile and a half (2.4 kilometers) in one direction stretched to over two hours, during which time we *looked* at rock formations and their components, *smelled* the differences between various types of pine trees, *tasted* leaves from multiple kinds of vegetation, and *felt* the textures of mosses and lichens that grew in the shade. Even though the park service technically has a natural sounds program, not once during the several hours that we sauntered to and from the lake did anyone mention any aspect of listening. And none of the hikers thought to ask.

During our walk, I heard several dozen commercial aircraft flying at around 30,000 feet (9,100 meters) heading from the eastern United States to San Francisco, Oakland, or San Jose. I heard people talking, their clothes rustling, and their cell phones—without which a few of the hikers apparently felt rudderless— ringing. And, of course, there was the sound of vehicles wending their ways through the valley somewhere off in the distance. I was also conscious of four different kinds of bird vocalizations: a pileated woodpecker (*Dryocopus pileatus*), an acorn woodpecker (*Melanerpes formicivorus*), a ruby-crowned kinglet (*Regulus calendula*), and a common raven (*Corvus corax*). There was only one vocal mammal, a gray squirrel (*Sciurus griseus*) sounding its alarm call, and no insects. Finally, there was the constant underlying natural ambient hush present at some low level throughout the valley floor, even though there was hardly any water flowing because of the drought, or wind blowing because of the abnormally calm weather. The components of the soundscape emanated from three different kinds of sources—human, non-human organisms, and non–biologically natural ones. The next day, when I asked those who had been on the walk if they remembered anything they had heard, only one person recalled the voices of a raven and the pileated woodpecker. None of the others remembered the impact of a single sound.

This culturally typical experience is one I've had many times throughout my journey as a bioacoustician—a collector and student of wildlife sounds.

I have a confession: because I grew up in a midwestern home completely devoid of animals, and in a family that cared little about wildlife, when I first wandered into a forest with a recorder and a pair of microphones, in 1968, I was scared to death and literally clueless about how to listen and capture the sounds of the woods. And it wasn't much of a forest, either. It was more like a park, an example of the "managed wild," a reasonably safe venue with well-groomed paths, lots of signage, and gift shops filled with things to buy. Still, it felt primeval and foreboding to me. I was on a blind mission—initially drawn there to hear and capture natural ambiences to include as part of an orchestration for an electronic music album and, in the process, find a location that produced something other than the din of human noise. There were no clearly defined goals.

No mentors. No teaching guides. No previous experiences to bring with me into the woods that day, ones equal to the task of capturing the sense of tranquility I imagined in my mind's ear. No one had introduced microphones and recorders capable of enduring the extremes of weather—and yet still able to record the subtleties of these sites. Protocols for taping whole natural habitats, whether on land or under the sea, were virtually unknown. Conceptual ways of perceiving and expressing most aspects of field recording were nonexistent, let alone the language for describing the phenomena revealed through sound. And there was no way of knowing, of course, the future revelation of the importance of natural sounds, and our human links to those miraculous voices. Humans had yet to realize the numinous epiphanies that would clarify the associations between the sounds produced by non-human organisms and our diverse cultures. In the late 1960s world of discovery, there was nothing *but* "future."

Taking a giant, well-timed leap forward, most everything about this topic still remains prospective, even today. My colleagues and I may have unlocked a few peripheral mysteries in the past decade, but we had then, and we have now, an astonishing amount of learning and explaining to do. In fact, the paths leading to these acoustic revelations and their connections to ecology are aglow with potential. The routes we choose to follow as a consequence may well set the stage for some of the most important decisions we will ever make.

For most of us, the acoustic world has always been an elusive one—an indistinct amorphic entity, unseeable and intangible—and listening is the "shadow sense."[1] Outside of musical literature, few words in English exist to explain the vast range of attributes that sounds express, especially in the emerging world of bioacoustics, the study of the sounds living animals produce. And yet, in different ways, sound insinuates itself into nearly every phase of our lives, its character having evolved from our primal connections to the natural world, to musical expression and language, to the preservation of sound in recorded media, and to the chaotic din we humans generate while navigating the course of our everyday rituals. Although we have rarely been conscious of those special bonds, remarkably, and unrealized until very recently, we are beginning to grasp that aspects of the soundscape inform disciplines as far ranging as medicine, religion, politics, music, architecture, dance, natural history, literature, poetry, biology, anthropology, and environmental studies. It is the exploration of these relationships that forms the basis for a new field of inquiry, *soundscape ecology.*

The combined sound sources experienced by the earliest humans, living in sub-Saharan Africa, re-sounded with complex signals. Some of those spoke to people spiritually. Others conveyed impressions of location, emotion, aspects of the hunt, and healing. Many inspired humans to break into song and dance. Early social groups rarely took these signals for granted, but as our lives became more urban-centered, the connections to those guiding beacons of the natural world began to lose their significance and consequently grew to be scarcely acknowledged. Since the onset of that progressive dismissal, several millennia passed before anyone stressed the importance of such sounds in Western culture. In the late 1970s, the Canadian composer and naturalist R. Murray Schafer coined the word *soundscape* to refer to the multiple sources of sounds that reach the human ear. Combining this with the word *ecology*, I use the resulting term, *soundscape ecology,* to describe new ways of evaluating the living landscapes and marine environments of the world, mostly through their collective voices.

This book addresses five basic questions, connected by a temporal thread, that explore the human bonds to our acoustic environment from the end of the Pleistocene to the present, and probes how this relationship will change in the future.

1. How did the term *soundscape* evolve?
2. How have soundscapes shaped human culture?
3. In what ways has technology influenced the future of the field?
4. How do different perspectives—human or other—reveal new trends in this discipline?
5. Through the lens of the soundscape, what future applications are possible or likely within several of those disciplines?

From the time we humans first occupied the forests and plains of Africa to our exploration of the most remote parts of the world, the sounds that surrounded us have engaged and influenced our imaginations. And because both academics and individual citizens have now acquired the technical means to capture and store sound, the future for understanding soundscapes is wide open. What I offer now are predictions unfixed in the moment.

I should note that this book is my personal view of the future of this important field. My hope is that this introduction will spur interested readers to investigate further. One place they might look is a book published in 2014, *Soundscape Ecology: Principles, Patterns, Methods, and Applications*, by Almo Farina, which digs into the details, including emerging definitions and field study protocols. Even highly detailed technical representations of this type show the promise of the field. Farina notes the use of soundscapes in applications as far ranging as the "assessment of the environmental quality of parks and protected areas, urban planning and design, ethology and anthropology, and finally in the long-term monitoring of the effects of climatic changes."[2] This book, of course, covers some of the same ground, but it is meant for the uninitiated, requiring no knowledge of acoustics or ecology—only an inquiring mind and a willingness to listen in new ways.

In assorted manifestations, notions of the sound-scape have existed from the instant we recognized the treasure of information emanating from acoustic expressions of the natural world. The soundscape concept consists of what I call signature sources, meaning that each type of sound, from whatever origin, contains its own unique signature, or quality, one that inherently contains vast stores of information. That individual signature is unlike any other. So, also, is the natural soundscape unique in its collective state, especially as it becomes the voice of an entire habitat. With my colleague Stuart Gage from Michigan State University, I have introduced new language meant to describe the three primary acoustic sources that make up a typical soundscape. The first is *geophony,* the non-biological natural sounds produced in any given habitat, like wind in the trees or grasses, water in a stream, waves at the ocean shore, or movement of the earth.[3] The second is *biophony,* the collective sound produced by all living organisms that reside in a particular biome.[4] And last is *anthropophony,* or all of the sounds we humans generate.[5] Some of these sounds are controlled, like music, language, or theater. But most of what humans produce is chaotic or incoherent—sometimes referred to as noise.

It is important to keep in mind the substantial impact each of these components may have on one another and how they interrelate. Each day that I peer into this universe, I feel the excitement that astronomers must sense when they discover a new far-off galaxy. As with astronomical findings, something new and surprising reveals itself in ways I would have never otherwise considered—revelations like the natural temporal and frequency bandwidth partitioning that occurs in wild soundscapes. *Partitioning* is a perfect descriptive term, one introduced in a paper by

Jérôme Sueur in 2001, which describes how cicadas carve out unique acoustical spaces where they can communicate without their voices being masked by others.[6] Some of my work has shown how the contours and botanical properties of the landscape itself can help determine how different organisms adjust their vocalizations to accommodate to those acoustic permutations.[7] While each of these iterations contains vital information we should not dismiss, to date few listeners have contemplated the inherent messages of those expressions, partly because of the growing distraction of extraneous noise in our environments, partly because natural soundscapes are disappearing, but also because of an increased human focus on visual events.

This book is a summary of my experiences; the stream of acoustic discovery that has been constant since I began listening attentively, recording, archiving, and writing about this field in the late 1960s. My hope is that it will encourage a new generation of active listeners.

In several places, this book refers to sound examples. I have indicated when a sound is available by including the symbol shown here. You can find and play these sounds by visiting the book's Web page, at www.yalebooks.com/soundscape.

1 ————————————

The Birth of the Soundscape

Surrounding our home in rural northern California, the sounds my wife, Katherine, and I have come to know consist of year-round aural traceries of birds, squirrels, amphibians, and insects in the mid-field, the personal conversations we share or the pleading voices of our cats, Barnacle and Seaweed, to be fed or released from indoor bondage to the wider world, the chatter of the TV or the whispered hum of the refrigerator compressor heard from the near field, and the sometimes irritating intrusion of commercial and light aircraft flying overhead combined with the far-off hushed drone of vehicular traffic from two miles (about three kilometers) away in the far field. The daily and seasonal sounds that define Wild Sanctuary, our home, convey a unique sense of place, one we've come to know as much by listening as by seeing.

The phenomenon of the soundscape usually consists of signals arriving from all directions on the horizontal plane and vertically from the sources overhead—a dome of 3-D sound and combinations of any or all of the three main sources mentioned earlier. Whether we're conscious of them or not, we're completely surrounded by acoustic elements coming at us from all directions. Active signals generally consist of biophonies and anthropophonies. Passive elements, such as wind and other weather-related signals, make up the rest. The impact of these sounds can be quite pervasive, depending on the environment. Sound, pressure waves transmitted through the air from a source to some type of receiver, can define the boundaries and structural properties of a room, a particular landscape. Of course, acoustic signals are also transmitted through marine environments and other media like wood and metal. The soundscape not only reveals the presence of vocal organisms that inhabit wild biomes, but defines the acoustic detail of floral and geographical features— think of the effects of wind in the trees or grasses, or water flowing in streams and by the lake or seashore. Soundscapes also expose the imbalance sometimes caused by changes in the landscape due to human endeavor or natural causes such as invasive organisms, weather, or movement of the land. One of my lifelong interests has been to find new ways to read, comprehend, and express these sources of information.

In 1939, a German ecologist named Carl Troll proposed using aerial photography as a measurement tool to determine relationships between ecosystems.[1] Landscape ecologists, practitioners of the interdisciplinary field that evolved from his work, study landscape structures and systems. Several decades later, in the late 1970s, R. Murray Schafer, with his colleagues Barry Truax and Hildegard Westerkamp, from Simon Fraser University in Vancouver, British Columbia, first defined the soundscape in ways that opened a window on an entirely new field of inquiry, and researchers have subsequently added to this prescient exposition.[2] Schafer's idea of the soundscape defines events as all the audio signals that reach our ears at any given time. The same goes for the acoustic receptors of non-human creatures. With the introduction of new descriptive language, such as geophony, biophony, and anthropophony, I was able to flesh out in greater detail the basic sources of sound. We have now cleared a path for a range of understanding in both science and culture that leads to fresh ways of experiencing and understanding the living world.

Even before the appearance of rangeomorphs, thought to be the world's first living marine organisms that evolved off what is now the coast of Newfoundland some 550 million years ago, there was still sound. But there was nothing living that was capable of receiving or responding to it. The wind still whistled and water still trickled along meandering paths. There was also the complex sound of volcanoes shaking the earth and filling the skies with ash. Then the sound of lava flows, landslides, glacial ice cracking and calving, ice in rivers and lakes breaking up in the spring, mudpots, and geysers. The extremes of weather-related conditions produced other acoustic dynamics, adding texture to the still unheard soundscape.

When organisms that could produce and receive sound first appeared, each type evolved to establish a clear bandwidth in the geoacoustic spectrum for its vocal behavior to be functional—these organisms needed sound-free channels in order to exchange vital information. At first, the sounds they produced were probably impulse-like signatures, more or less like static or short pops and cracks meant to be loud enough in amplitude that they could be sent and received above the levels produced by the surrounding geophonic ambience. Although tiny insect-like marine organisms have developed to generate some of the loudest sounds on the planet (relative to their size), researchers do not generally think that the earliest vocal organisms could communicate beyond a range of a few centimeters. As organisms grew more sophisticated, numerous, land-based, and complex, their sound-producing and receiving capacities also became more efficient. Unique species created unique acoustic expressions, acoustic signatures. Crickets stridulate by rubbing their wings together—one with a kind of file-type apparatus and another with a scraper. Fish resonate their air bladders, crunch at coral, or oscillate their caudal fins at rates that generate underwater signals. Crabs snap exoskeleton claws. Ants either strike branches with their bodies to transmit sound signals throughout a tree or they stridulate by rubbing their legs on their abdomens. Barnacles twist in tiny shells, birds push air through syrinx-like organs, snakes expel air quickly to hiss, dolphins project highly focused signals through melon-shaped organs, kittens and lions vibrate muscles in their throats to purr and roar. But when living organisms became more numerous and began to fill acoustic niches in their respective habitats, their voices necessarily had to adapt through partitioning, so that each one could transmit and receive signals unimpeded in the specific time or range necessary for their survival. Efficient uses and conservation of energy were paramount.

Some of these observations can lead to astounding correlations. For example, thunderstorms that occur with great frequency in the equatorial regions of the planet give off electrical signals that are then transmitted through the earth's magnetic field to the north and south poles. Low-frequency radio receivers such as those used by NASA can detect and record these signals. When played back, they sound like long, extended ascending and descending glissandos produced in early analog synthesizer performances of the mid-1960s. There is nothing particularly unusual about those phenomena. What *is* curious are the remarkably similar vocalizations produced, in turn, by Weddell seals (*Leptonychotes weddellii*) in the Antarctic, and bearded seals (*Erignathus barbatus*) some 12,000 miles to the north, in the Arctic. Their respective vocalizations not only closely replicate the signals detected by the NASA radio receivers, but they're similar to each other even though the two species have never met.

Though such claims still need substantiation, some animal physiologists, like the late Frank Awbrey, a marine biologist from San Diego State University, have suggested that a small amount of magnetite in the heads of both species may give them the capacity to re-ceive these electronic signals and imitate them, much like the anecdotal stories of people picking up faint radio signals through the silver (amalgam) fillings in their teeth.

During the Pleistocene, when humans first popu-lated African habitats, we added our range of voices to the biophonic mix. As skillful mimics, we first learned to emulate the complex roars, grunts, rhythms, melodies, and harmonies that we experienced in our respective habitats. The mimicry helped us to play it safe; there was little room for error in our attunement with life around us. It was the closest we came to actual biophonic expression during the course of our evolu-tion. As a small fraction of the larger biological com-munity, we owed tribute and calculated the benefits of deference. The soundscapes of the forests and plains signaled where food was (or wasn't). They provided a necessary voice of the divine that, in turn, delivered answers to ontological questions we otherwise had little capacity to solve. They inspired us to organize sound into complex patterns, reflecting those heard in natural settings, which culminated in the first expressions of music and dance, and probably even language. They nurtured us with pulses and cycles of reassuring sonic textures that calmed and centered us—that confirmed our place in the living world. And the biophonies also served as an aural GPS, signals that guided us to remote locations under the cover of total darkness and through the densest foliage with extraordinary precision.

Stumbling through the forest on a nighttime hunt with the Jivaro—a group living in the Amazon Basin in a large area that encompasses eastern Ecuador and part of Peru—I was awestruck by the accuracy with which sound guided hunters along their routes. It was not the beam of a flashlight, the glow of a torch, or the faint light, obscured by the forest canopy, of our moon or distant constellations overhead that guided the Jivaro, but amoeba-shaped grids of biophony through which they walked. The acoustic partitioning of a wide variety of insects and amphibians constantly updated the hunters with salient information about their respective location at any moment, the potential objects of their hunt, and the direction their game headed even when they were unable to see any trace of it several hundred meters from the group.

Until such experiences in the early 1980s, I had pretty much dismissed wild soundscapes. But what I learned from this and many subsequent encounters was that careful listening gives us extremely valuable tools with which to unlock edifying codes, ones extending way beyond the narrow limits we have typically extracted from human-produced acoustic media. Wild soundscapes, the voices of the natural world, provide exceptionally instructive perspectives through which to connect with the living planet. They are nothing less than markers that direct us to places of refuge and narratives that can point us to avenues of healthy survival. To ignore them is to deracinate from our lives the numinous thrum of our very existence.

My archive of environmental sound, collected over nearly half a century, represents many heterogeneous soundscapes of both marine and land-based environments. It exceeds five thousand hours of holistic habitat recordings, and in excess of fifteen thousand species. That is not a lot of data by today's standards. As I will describe, with current digital technologies and arrays of multiple recorders, researchers can easily capture many times more data in just a single week or month from one location. What is notable is that much of that collection before the mid-1980s was, of necessity, recorded in analog formats. Keep in mind that these recordings were made with heavy, cumbersome field recorders that used seven-inch reels of quarter-inch wide (.635 centimeter) audiotape. In the early 1980s, Sony introduced the first digital recording system, the F-1 format, which recorded data onto a VHS (Video Home System) cartridge-like package. And in 1987, with the introduction of a transitional medium called DAT (digital audiotape), the large F-1 cartridge evolved into a much smaller unit. Because these formats were electromechanical and had moving parts, they sometimes encountered operational problems caused by humidity, temperature, or rough handling. In addition, the limits of the digital audiotape medium itself constrained the amount of continuous time we could collect sound in the field unless we had the capacity to carry large quantities of batteries, tape, and backup recorders.

A bit late responding to the technological change, I converted to purely digital media, hard drives and compact flash, in 2004. By then, I was beginning to slow a bit physically and missed some of the great new wave of recorders and microphone systems, as well as the opportunities to use them, a hesitant but calculated retreat. As of this writing, digital field recording technologies can store up to eight tracks on professional-quality devices roughly the size of a paperback book. It is not unlikely that we will soon see many more microphone or line inputs added in the future along with smaller-scale recorders.

As field recording systems have become priced and designed to be more accessible, easy to use, and functional in the face of climate extremes in the field (devices are now suited for the range of humidity and temperature conditions present in arctic, equatorial, or desert regions of the world), I expect that an increasing number of individuals and groups would aim to record in ever more wild habitats, with the caveat that those sites are visited lightly and no damage or residual evidence of our presence remains.

In spite of the vast quantities of data we can now collect in a short time, the most significant aspect of

the archive stored at the Wild Sanctuary is the heart-
breaking fact that more than 50 percent of the material
recorded over nearly five decades comes from sites so
badly compromised by various forms of human intervention that the habitats are either altogether silent or
the soundscapes can no longer be heard in any of their
original forms, otherwise revealing dysphonia—a word
that in medical terms means an inability to speak.
It's one that translates well to this field. By itself, this
baseline collection is so rare, especially the recordings
made in North America, that it might be considered on
the same level of importance as other national trea-
sures. From a global assessment, dysphonic wild habitat trends clearly point to an increasing and alarming
rate of loss of the voice of the natural world.

Two examples from my own experience point to
some of the unanticipated effects of human endeavor.
The first of these took place at Lincoln Meadow, about
a three-and-a-half-hour drive east-northeast of San

Francisco at roughly 6,500 feet in the Sierra Nevada. I had recorded there for many years, and in 1988 a timber company tried to convince local residents that there would be no environmental impact from "selective logging"—harvesting only a few trees from a site instead of clear-cutting. Operating on a hunch and with permission granted to record before and after the event, I captured representative samples of what the soundscape was like prior to the operation. I could clearly hear and see (by examining a visualization of the frequencies present, called a spectrogram) a stream running through the meadow and many diverse bird signatures (fig. 1).

A year later, after the logging operation, I returned and, using the same protocols, captured more examples. Listening this time, the stream sounds were still there, but the rich biophony, present in all previous years, was now practically silent (fig. 2).

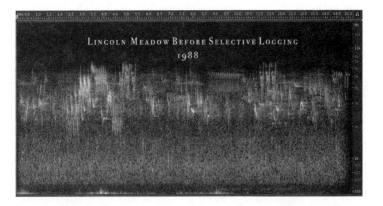

1. Spectrogram from Lincoln Meadow, June 1988, biophony before selective logging

2. Spectrogram from Lincoln Meadow, June 1989, biophony after selective logging

I've returned to that site fifteen times in the past quarter century, but the density and diversity of the biophony has not yet recovered. To the human eye and to a camera, the landscape looks unchanged and would have supported the logging company's sustainability contention. But the soundscapes tell a very different story.

In Costa Rica, where I had recorded in the Osa Peninsula beginning in the late 1980s, I encountered the effects of clear-cutting. At one location, in 1989, I recorded at a site that consisted of many thousands of acres of old-growth forest (fig. 3).

When I returned seven years later, in 1996, that whole section of forest was covered with a grid of logging roads and many sections had been clear-cut, especially the site where I had recorded in 1989 (fig. 4).

The before-and-after comparison for each of these

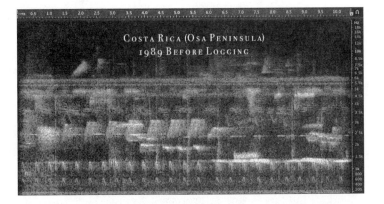

3. Spectrogram from Osa Peninsula, Costa Rica, 1989, before clear-cutting

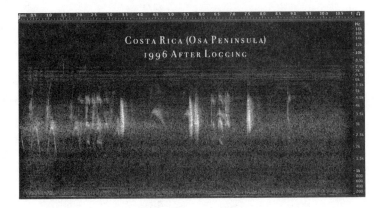

4. Spectrogram from Osa Peninsula, Costa Rica, 1996, after clear-cutting

two sites, Lincoln Meadow and a location in the Osa Peninsula, was based on intuition in the first case, and purely accidental but not unexpected in the second. One of the major contributions to the field of sound-scape ecology will be precisely these types of assess-ments, planned from the outset with careful attention to calibrations of microphones, recorders, input and output levels, and appropriate metadata such as weather, date and time, microphone placement, and GPS information (longitude, latitude, and elevation).

The recording of natural sound did not begin with the holistic idea of capturing entire biomes, typically wild habitats. For one thing, microphone and re-corder technologies simply weren't robust enough. The pursuit of recording wildlife sounds began, instead, partly by accident with a model that paralleled a firmly entrenched nineteenth-century approach to science—one that focused on the abstraction, deconstruction, and fragmentation of a single species out of the context of its respective environment.

Even though the technologies to record and reproduce holistic audio performances began to materialize around the time I started on this quest, "recordists" interested in the natural world did not, for the most part, contemplate soundscapes as significant models for collection. Except for limited use in film and television sound tracks, whole-habitat recordings were virtually unheard of. Rather, a reductionist view of the acoustic world that embraced fractured and incongruous acoustic signatures—distorted snapshots of solo animals in a kind of bioacoustic zoo—remained the dominant field-recording ideal. Throughout much of the twentieth century, those of us in the field were charged with carefully abstracting brief individual sound signatures from within the whole acoustic fabric, splitting them apart into unrelated morsels.

The inspiration for sound fragmentation came

about, in part, because several ornithologists were curious to learn if there was a way to capture individual bird and mammal voices separated from the context of the larger soundscape, likely considered by them to be indistinctive and extraneous noise. Their initial curiosity may have been to push the limits of the technologies they were using rather than a conscious splintering of phenomena in the natural world. However, these efforts resulted in a far different outcome. The researchers clearly found themselves on an assignment. After discovering that they could isolate the sound of a single bird by using what the recordists of the 1920s called a sound mirror—an early version of the parabolic dish—and recording the signal to the optical track of a Movietone sound recorder originally designed for film, Arthur Allen and Peter Paul Kellogg set out from the Cornell University Lab of Ornithology to chase down and record the rare ivory-billed woodpecker. In the spring of 1935, after they finally spotted the bird and its nest, the researchers captured one clear recording of the creature, which had been imagined extinct.[3] (More recently, since February 2005, a few sightings of the bird have been confirmed in the Big Woods of Arkansas.)

Meanwhile, across the Atlantic another audio naturalist, Ludwig Koch, had migrated from Germany to the United Kingdom to work for the BBC in 1936. From the late 1880s, he had been recording individual birds both on the Continent and subsequently in Great Britain, later using a device more akin to a Vitaphone—a disk recorder in which the needle records and plays back the sound, tracking grooves from the center of the disk to the outer edge. The decontextualized single-species model that Koch and North American wildlife birders established ushered in a narrowly focused academic format still favored in many circles almost a century later. Based on the idea of *life lists*—finding and identifying single species of birds and mammals and, more recently, amphibians, reptiles, and insects—the approach of collecting animal sounds by the numbers became firmly entrenched while curiosity about the larger vocal collective was met with an attitude that was utterly phlegmatic.

Because listening to only a single species had been the standard mindset for nearly the entire history of audio recording, others entering the field followed the same path or risked ridicule and even outright derision. The focus on single sound fragments initially forced both casual listeners and serious researchers to confine their inquiries to the limits of each vocalization, whatever its origin. Of course, there were many justifications for this approach, one being that isolating the calls helped researchers determine if the vocalizations were language-like and to understand how other individual organisms created and perceived calls. Yet I was always troubled by the strong feeling that this older, piecemeal acoustic model distorted a sense of what was wild by giving us an incomplete perspective of the living landscape. Moreover, it was my sense that the answers researchers were looking for in the single-species paradigms might have been more easily revealed by considering the context of those expressions from a more holistic perspective. Although those questions still need to be resolved, my guess is that, until now, we've essentially ignored a necessary link between the human and non-human aural phenomena as a consequence.[4]

Undaunted by those impediments, one of the most thrilling aspects of my work has been the discovery of the *niche hypothesis,* an early stage of the biophony concept—the collective and structured sound that whole groups of living organisms generate in a biome at any given moment.[5] The term, originally proposed by my colleague Ruth Happel, became clear between 1983 and 1989 and led to the observation that the makeup of wild soundscapes was primarily a form of expression where each type of organism evolved to vocalize within a specific bandwidth—based on either frequency or time. That, in turn, shed light on the bioacoustic relationships between all of the organisms present in a particular biome. In other words, in order to be heard, whether in urban, rural, or wild habitats, vocal organisms must find appropriate temporal or acoustic niches where their utterances are not buried by other signals.

More recently, we have begun to explore the interactions between biophony and the other sources of sound: geophony and anthropophony. For example, several studies, particularly those in process by Nadia Pieretti at Urbino University in Italy, have shown that birds alter their vocalizations to accommodate themselves to urban noise.[6] And killer whales (*Orcinus orca*) do the same with boat noise in their marine environments.[7] Other studies have led to the observation that the music and language produced by a few remaining indigenous cultures can be directly tied to the intricate soundscapes of the natural world that defined their respective habitats, in a rare instance of human bio-phonic expression.[8]

Gary Snyder, the American eco-poet elder, has pointed out that human language is wild, organizing and reorganizing itself independently of human will.[9] My efforts have shown that, in much the same way, the communicative structures in certain undisturbed biomes form a basis for that paradigm, a constantly changing, reflexive synthesis of correlated sound and its subsequent harvest. Our current understanding of the natural world soundscape necessarily pinpoints "acoustic niches," the special ways different species in a single soundscape use to jostle for sonic territory. By recognizing the function of this partitioning, or forma-tion of acoustic and temporal niches, a creative and important realization emerges: that soundscape ecol-ogy is no less crucial than spatial or landscape ecology for our understanding of ecosystem function. Animal communication turns out to be as significant a factor in defining material or acoustic real estate, habitat, and ecological integrity as, say, trophic structure—the feeding and nourishment relationships of all organisms in a specific environment. In fact, territory, habitat, and ecological integrity may no longer be broadly definable in three spatial dimensions alone. The addition of soundscape ecology adds a fourth.

With our cultural focus primarily on visual experience and manifestation, we seem to have lost the delicate balance informed by incorporating all of the senses in our awareness of place. Nevertheless, it is imperative that we engage with wildness through its multiple dimensions. In that way, our inclusion of the holistic acoustic model enlarges our sense of the wild by literally expanding the boundaries of perception. It also rivets us to the present tense—to life as it is—singing in its full-throated choral voice and where each singer is expressing its particular song of being. It is my goal to encourage us to take a deeper plunge into a wilder world beyond the mundane and merely visual, suggesting that the natural wild is both more complex and more compelling than meets the simple eye. As I always remind my students, "A picture may be worth a thousand words, but a natural soundscape is worth a thousand pictures."

2 ───────────────────────────────

The Challenges

I t is clear that humans, as mimics, integrated sound
from various natural sources into numerous cultural
modes of expression. But when I began to think about
this idea, the reverse perspective became more inter-
esting to me: that is, how have wild soundscapes
accommodated the far-reaching effects of anthropophony as a part of its continuously evolving fusion?
Moreover, in what ways have living organisms—human
or Other—experienced soundscapes? Other, in this
context, includes all organisms that are not human.[1]
There was a time, for example, maybe as recently as fifty

thousand years ago, when human sound did not domi-
nate the landscape. Human expression, from whatever
source, wasn't as loud as thunder, or the impact noise of
a falling tree in the forest, an avalanche, the explosive
boom of fifteen- to forty-foot (4.5 to 12 meters) waves
in a storm, the pitched effects of raging wind, volcanic
eruptions, or the attendant rumble of earthquakes.
Many natural sources were louder than anything humans felt they had the need to reproduce or exceed.

In the beginning, and aside from other multifaceted
connotations, the biophonies and geophonies of the
forest likely spoke to us as "spirit voices," mysterious
utterances unexplained by any known logic of those
emergent times. Because many of us have a deep-
seeded need to provide answers for everything, we
relegated those phenomena to initial concepts of the
divine. Known as *boyobé* to the Ba'Aka—pygmies who
live in the Dzanga-Sangha rainforest in the remote
southwest region of the Central African Republic—the

dimming light of the crepuscule hid multiple spirit forms outfitted in fantastic masks and costumes that would spring from behind the shadows of trees and bushes to join festivities of singing and dancing. But first, the "spirits" had to adorn themselves so that the rest of the group members would consider them invisible. At the same time, while embracing the primary biophonic signals of the surrounding forest, the other group members were inspired to perform, welcoming the veiled essences with the characteristic grace of dance and song. The ability to imitate movement and sound with requisite esteem is seminal to this reverential act. In fact, the Ba'Aka, just as we once did, learned not only to impersonate individual sounds, but to incorporate into their music the expression of the soundscape as a whole—its remarkable and eloquently defined moment.[2]

In the process, we also emulated the movement of other living organisms. This was done in the form of dance—another effort to plumb universes beyond our own. This combination of "musicking," to gain from our environment a deep sense of the divine, calmed and helped heal us from the stresses encountered in everyday life.[3] They are similar to the responses of the Ba'Aka, who disappear into the rainforests to regenerate spiritually, emotionally, and physically, after the pressures imposed on them from contact with the cash economies of the West and East. Anecdotal evidence strongly suggests that natural soundscapes and biophonies, in particular, may lower stress indicator levels (glucocorticoid enzyme, heart rate, blood pressure, and so on) in humans far more successfully than environments saturated with music, because music is culturally biased and may actually produce a result opposite from what was clinically intended or hoped for.

To some, however, the simple term *soundscape* can be intimidating. And anyone thinking about the future of this field should be aware of the challenges.

Despite its relevance to human culture, *soundscape,* a derivation of "landscape," or "seascape," has had its detractors. The National Park Service implemented a well-vetted soundscape program in the early 2000s, with the express goal of encouraging visitors to hear and engage with the natural sounds around them. But the agency had no sooner incorporated the concept into several of its visitor and internal resource management activities when two members of the U.S. House of Representatives challenged its origins, implicitly threatening to target funding for the park service if the term wasn't changed. Republican representatives Don Young and Richard Pombo, from Alaska and California, respectively, wrote a letter to Gale Norton, head of the U.S. Department of the Interior under George W. Bush, in 2003. The letter was in response to noise issues raised by the park service with the Federal Aviation Administration relating to tourist flights over sites like the Grand Canyon and other parks, creating a serious problem for visitors and wildlife alike. After asserting that the term *soundscape* had not been sufficiently characterized or established, the letter said, in part:

> the concept of "natural soundscape" has never been clearly defined and continues to be applied in the most extreme way. Therefore, we ask for you to respond to each of the questions below:
>
> . . . What is a "natural soundscape" and how does it apply to the evaluation of noise impacts to parks? . . . As a new concept, has the term "natural soundscape" ever been subject to public notice and comment? . . . Will the "natural soundscape" policy ultimately lead to the banning of human activity in, near, or over the parks? . . . We want to be clear that the "impact" that the purported "natural soundscape" policy could have on the national aviation system is unacceptable.[4]

Young and Pombo's letter to Norton resulted in Fran Mainella, then director of the National Park Service working under Norton, being strongly "encouraged" to impress on the soundscape office in Fort Collins, Colorado, the need to alter the word. In 2004 the Natural Soundscapes Program name was changed to the Natural Sounds Program, which remains the designation today. That might seem a minor change, but the result of this dispute was not just semantic: many of the visitor soundscape programs assembled in the late 1990s and early 2000s were subsequently abandoned, broadening the impact of this language revision.[5]

A similar example of government intercession, to suppress the outcome of a noise study, took place while I was working on my doctorate in the late 1970s. Engaged as a marine bioacoustics intern in research approved and financed by the National Park Service in Glacier Bay, Alaska, my group hoped to determine why, despite ample food resources, humpback whale (*Megaptera novaeangliae*) population numbers in the bay were declining. Researchers reported that humpback whales appeared to be swimming away from the irritating sounds of large tourist vessels that generated huge amounts of propeller cavitation, the formation of bubbles as a result of its movement through liquid, and engine vibration. The whales, in response to the harsh noises, hid in the acoustic shadows of island landmasses or large bodies of calved ice. The report concluded that uncontrolled loud vessel noise contributed to at least one of the major probable causes of the whale population decline in the bay.

For several years the report was not made public because, according to Charles Jurasz, a biology teacher and naturalist from Juneau and the principal investigator of the study, James Watt, then Ronald Reagan's secretary of the interior, ordered the National Park Service to quash the findings.[6] The study might have had a negative impact on tourist-vessel traffic in the bay, and the park service was apparently compelled to comply with the order. Jurasz, an environmentally engaged local activist, was never again able to obtain the necessary permit to continue his work in Glacier Bay, to verify his data or do a follow-up study. The rejection left him devastated and bitter. Jurasz, whose groundbreaking observations of humpback whales had helped us understand how the animals feed, was the first scientist to describe how humpbacks dive down tens of meters deep, swim in circles blowing bubbles to force large numbers of krill into a compressed ball, then lunge up from below in this "bubble net" with their maws wide open to capture a meal. The National Oceanic and Atmospheric Administration recently honored Jurasz for these and other findings.[7]

Over the past few decades, redesigned vessel engines, hulls, and propellers configured to generate less vibration have helped to mitigate the noise emitted by commercial vessels in Glacier Bay. Recent reports show that the whale populations have returned to "near normal" numbers. Allison Banks and Chris Gabriele, park service employees at Bartlett Cove in Glacier Bay, reported in June 2000 (and again, through personal correspondence, in 2010) that the humpback whales are again thriving.[8]

During the past five decades, noise in the U.S. national parks has been an ever increasing problem. The mandate under the National Park Service Organic Act of August 25, 1916, states that "the Service thus established shall promote and regulate the use of Federal areas known as national parks, monuments and reservations . . . by such means and measures as conform to the fundamental purpose of the said parks, monuments and reservations, which purpose is to conserve the scenery and the natural and historic objects and the wild life therein and to provide for the enjoyment of the same in such manner and by such means as will leave them unimpaired for the enjoyment of future generations." In spite of these clearly stated objectives, there have been escalating tensions as special interest groups weigh in with claims of extraordinary rights. In February 2002, the National Park Conservation Association invited me to do a preliminary snow- mobile noise evaluation in Yellowstone National Park. Forceful constituencies were demanding unlimited use of snowmobiles in the park, and they were sup- ported by numerous U.S. lawmakers. Aside from the atmospheric pollution generated by the engine exhaust from snowmobiles, these vehicles have been produc- ing increasing amounts of sound pollution—chaotic signatures of broad-ranging loud noise that encom- passes much of the human and non-human-creature frequency spectrum, and carrying several miles from the main roads into territory that would otherwise be quite tranquil during the winter months. Our study was one of many designed to quantify the effects of noise.

A groundbreaking paper by Scott Creel, published in 2002, evaluated the effects of snowmobile noise on wolves and elk in Yellowstone, Voyageurs, and Isle

Royale National Parks. It shows direct correlations be-tween increased snowmobile noise at those venues and high levels of stress indicators in the animals.[9] With straight-piping motorcycles, jet skis, dune buggies, dirt bikes, and ATVs infusing the park habitats with their raucous sound signatures, those all-encompass-ing full-frequency noise emissions affect everything from the tranquility visitors expect to the measurable stress caused on wildlife. The strain on wildlife and humans as a direct result of anthropophony remains a problem in spite of reduced numbers of vehicles and park monitoring.

In 1982, a year after Ronald Reagan came into office, he appointed James Watt as secretary of the interior. One of Watt's first acts was to defund the Environmen-tal Protection Agency's Office of Noise Abatement and Control (ONAC), the sole federal agency charged with promoting a quieter, less harried America. Although the office is still technically in place, the loss of funds essentially shut down the program. The office was formed under the authority of the Noise Control Act of 1972, which is theoretically still in force, and the ONAC officially remains part of the federal government, but its responsibilities have been shifted to the Department of Housing and Urban Development and the program continues to be understaffed and rendered powerless, with little or no enforcement authority. Despite several attempts in Congress to resurrect the agenda through the introduction of bills such as the Quiet Communities Act of 2003, none have been successful. In response to a student's question of why he shut down the ONAC, Watt said: "Noise is power. The noisier we are as a country, the more powerful we appear to others."

Researchers seek answers to questions like how to articulate the effects of climate change, or resource extraction in a given habitat, or the effects of ocean warming and acidification on a coral reef, or the effects of anthropophony on a given habitat's biophony. The biggest problem bioacousticians face is how to generate study models that express what we are finding in the field through our recordings. The major difficulty that field-recording scientists have to overcome is the generation of solid baseline data to use in comparing site changes over time.

Michigan State University's Envirosonics lab and REAL (Remote Environmental Assessment Laboratory), the bioacoustics programs at the Muséum national d'histoire naturelle in Paris, and the Department of Basic Sciences and Foundations at the University of Urbino continue to publish advances in measuring these effects. Researchers at these institutions have generated accepted models, now applied across much of the resource management community. At the same time, there are still holdouts who dismiss these findings in many industries, such as mining, forestry, and oil and gas. Driven by specific goals and agendas, this group typically features those who have weighed in on these matters by supporting and endorsing industry-friendly political candidates typically known for disregarding compelling scientific research and accessing natural resources with as little government oversight and restriction as possible.[10]

Recording and analysis of biophonic data allows for a very rapid, efficient, and precise means of evaluating bioacoustic density, diversity, and richness. It is information directly related to habitat health, the cause and effect of resource extraction, land transformation, global warming, and natural geologic and weather-related events. Each recorded example of habitat sound-scapes (biophonies) is a representation that instantly clarifies the status of any given biome in terms of its relative fitness. All of this information and these types of approaches are precisely the kinds of data many in the resource extraction and industry-indebted political communities would prefer not to become transparent or widely known. The most obvious reason for such resistance is that, so far, there has been no language or format for rebuttal except for frustrated utterances of strongly worded skepticism.

Given these obstacles, I have to constantly keep in mind the goal of helping to maintain a healthy, living planet, one that sustains all life. Natural soundscapes fluently reveal the scorecard. More answers probably reside in a question that the author and journalist Richard Louv has asked about our progeny: "Are we going to be the last generation of children who grew up playing, learning and thriving in the woods?"[11]

3 ⎯⎯⎯⎯⎯⎯⎯⎯⎯⎯⎯⎯⎯⎯⎯⎯⎯⎯

Technological Progress

Around 2000 BCE, when the first musical notation appeared, making possible the reproduction of certain types of acoustic performances in real time, humans had taken the initial steps toward capturing and preserving musical presentations. But until about 150 years ago, there was no method to capture actual sound in the form of a recording. That changed when Édouard-Léon Scott de Martinville, a French inventor, recorded the first known acoustic impressions represented as lines on smoke-blackened paper or glass in 1857. Following on his heels two decades later, Thomas

Edison invented a vastly improved mechanism that, unlike de Martinville's device, could instantaneously reproduce what had been recorded. No one at that time could have foretold what would eventually happen in the field of music or natural sound as a result of Edison's invention. In 1889, the German naturalist Ludwig Koch made the first known bird recording of a common shama (*Copyschus malabaricus*), when he was only eight years old.[1] He was also one of the first to use a primitive version of the parabolic dish, a device with a half dome about a meter in diameter and a microphone mounted in the middle facing backward into the center of the dish (fig. 5). Having set out to capture birdsong mostly, later in his life he added a few mammals and amphibians to his collection.

This particular technique—the incoherent recording of creature voices out of context—became the leading sound capture paradigm for over a century and, to some extent, is still favored. The notion is fueled

5. A parabolic dish microphone, from Telinga Microphones

by the belief that one learns the behavior of certain species from the inside out and not as an inherent part of a larger structure. From the perspective of most institutional thought, this has been the primary way to learn the individual expressions ascribed to a single organism.[2] To me, the approach has always seemed a bit like trying to understand the magnificence of Beethoven's Fifth Symphony by abstracting the sound of a single violin player out of the context of the orchestra and hearing only that one part. From a technological standpoint, the good news is that, with new field recording technologies and even with redesigned stereo parabolic microphones, we can both accomplish the capture of entire biomes *and* retain the ability to separate out individual voices from the whole soundscape simultaneously.[3] Add to that the power and discriminating abilities of currently available editing software and it is possible to hear and view the context in which most organisms vocalize in relation to all the other animals in their habitat, while at the same time benefiting from the flexibility to abstract a single voice and hear it separately. So the rationale for the less instructive types of acoustic deconstruction almost exclusively supported during the past hundred years no longer enjoys the same exigency.

From the time of their invention through the first decade of the 1900s, wax cylinder technologies for reproducing audio were dominant. Strictly a mechanical device, the audio track was engraved on the outside of the cylinder with concentric grooved lines. When someone either played or sang very loudly into the wide end of a large flared horn, the vibrations were picked up at the narrow end by a diaphragm with a needle attached. As the cylinder rotated somewhere around 160 times per minute, the needle traced squiggly grooves onto the wax surface, transforming them into representations of an audio signal. When played back, the needle retraced the grooves, causing the diaphragm to vibrate accordingly. The vibrations generated by the needle's contact with the grooved surface of the cylinder were transmitted to the diaphragm, the signal of which was then amplified through the same (or a similar) horn.

By 1910, phonographic disks had begun to replace the earlier cylinder technologies. Flat "records" had the advantage of being able to capture signals on both sides, they were easier to manufacture, store, and ship, and they also reproduced sound more faithfully. By 1927, single-sided cylinders were a thing of the past. Meanwhile, the disks quickly advanced from 78 revolutions per minute (RPM) through 45 and later 33 RPM, and the vinyl medium replaced the shellacked 78s soon after World War II. "High fidelity" recordings, especially after the introduction of stereo sound in the 1950s, reproduced sound more authentically than ever, and they were also easier to reproduce in large quantities than previous technologies. These records also mitigated the issue of breakage during shipment and they were easier than the older 78s to store in large quantities.

Clément Ader, a French inventor and engineer, conducted the first experiments with stereo transmission in 1881 at the Exposition internationale d'électricité (International Exposition of Electricity) in Paris. An array of early transducers (primitive microphones) was arranged on either side of the lip of the stage at the Paris Grand Opera. Cables ran from there to a site at the Palais de l'industrie (Industrial Palace), three kilometers (just short of two miles) distant.[4] Although I had known about the multichannel and stereo recording of *Fantasia* in 1940, with Leopold Stokowski conducting the Philadelphia Orchestra for the music track to the Walt Disney film, I didn't know until recently that Duke Ellington and his band had recorded a stereo session as early as 1932.[5]

Just as many in the recording industry focused on replication of music and a bit of spoken word, except for a few species-specific recordings almost no one was capturing the natural soundscapes of whole habitats. Even more problematic was the difficulty of operation, transportability, and capture and reproduction capacity of early recording technologies. There were also no established field recording models for acoustically reproducing the more comprehensive aspects of the natural wild.

It wasn't until the portable stereo tape recorder became available in the late 1960s that the idea of capturing natural soundscapes from a more holistic perspective occurred to a few of us, like Walter Tilgner in Germany, Jean Roché in France, Chris Watson and Martyn Stewart in the United Kingdom, Murray Schafer and Dan Gibson in Canada, and me in the United States. When these machines were paired with two matched microphones connected to separate recording channels, it became possible to collect coherent sound information that reproduced the illusion of space and depth. Even with this vastly improved sound capture and reproduction, however, the equipment was still unwieldy. The best recorders from the late 1960s to the mid-1980s, like the Nagra IV-S, for instance, weighed nearly thirty pounds (approximately fourteen kilograms) and were powered by a set of twelve D-cell batteries that lasted only about five hours before they needed to be changed (fig. 6). Seven-inch (17.78 centimeter) reels of audiotape used on the Nagra recorders weighed nearly a pound (almost half a kilogram) and could record only for a bit over twenty minutes per reel at a speed of fifteen inches (38.1 centimeters) per second, the normal velocity of professional field recording equipment of the time. And that didn't include the additional weight and bulk of microphones and mounts, headphones, tripod, cables, spare batteries, and extra reels of tape.

Field recordists, at the onset of natural soundscape collections in the late 1960s, needed to be an intrepid lot; in addition to being young and physically agile, they also had to be inventive, since there was absolutely no written source for picking up useful techniques. It was all learn-as-you-go. It took me a full

6. Nagra IV-S analog tape recorder, Sony TCD-D10 digital audiotape recorder, Sound Devices 722 digital recorder, Zoom H4n digital recorder

decade, for example, to figure out how to record the sound of ocean waves so that it sounded like the real thing on playback. It is a lesson I could now impart to a soundscape ecology intern in less than five minutes.

The technical models began a rapid transformation in the 1980s with the dawn of intermediate technologies that were quickly shifting from analog to digital formats—from awkward and heavy reels of audiotape to hard drives and large-capacity flash memory cards. The first of these digital devices was the Sony F-1, which ultimately proved to be too cumbersome for field applications. DAT recorders, at the time of their commercial introduction around 1987, were much lighter and smaller than their analog and digital antecedents; they were about the size of an average hardcover book. They still depended on small tape cartridges, but the audio information picked up by the microphones was converted into digital data, which, in turn, was stored on tape. An improvement over previous analog systems, the DAT recorders used only two or four D-cell batteries for the larger systems. During the late 1980s and early 1990s, even the DAT transports themselves had shrunk to the point where some of the recorders were not much larger than a deck of playing cards and could be powered by two to four AA batteries. This reduced the weight of the field gear by more than 80 percent, compared with the earlier technologies. The recorders were also much more energy-efficient.

In addition, DAT technologies increased the total data storage time on one cartridge by more than a factor of three—triple the recording time without having to change tapes. By the late 1990s and early 2000s small digital disk recorders were introduced as other transitional pieces of gear. But it wasn't until large-capacity hard drive and compact flash recorders appeared around 2004 that it became possible to capture very high quality sound using increased sampling and bit rates, along with the ability to record continuous audio segments lasting for dozens of hours.

Microphone systems, too, have greatly improved. For my work and for simplicity's sake I have divided microphones into three categories: monaural (a single-source technique), stereo (any two-source array of closely aligned microphones, as expressed through many subcategories), and multichannel systems (with more than two inputs).

At the time I began recording in natural unchanged habitats, microphone systems were meant to be principally used in temperature- and humidity-controlled environments like concert halls and recording studios. So when I first tried them in a forest environment, even temperate ones, condenser systems almost invariably failed. The slightest exposure to humidity caused the sensitive capsules to fail. And, because of their cardioid patterns (the heart-shaped detection pattern of acoustic sources in this case), they were ultra-sensitive to wind noise, making it nearly impossible to record in all but the most sheltered locations. There weren't many microphone options except to switch to the less sensitive and (to me) more limited dynamic types, which I always carried as a backup even though they were heavy and awkward to use. In addition, the dynamics were typically noisy because of the large amount of amplification required, and did not reproduce the feeling of depth and width that other condenser mic technologies delivered. Now, instead of an extra set of dynamics, my backups usually consist of one or two pairs of electret microphones—not terribly sensitive to humidity. Given the increasing interest in wild ambient recording in the field during the late

1980s, Sennheiser, a microphone producer in Germany, began to actively promote a branded condenser format that could withstand highly humid conditions for long periods of time without failure. Several manufacturers have since incorporated this type of approach into their designs.

By 2010 field technologies had improved to the point where pairs of omnidirectional stereo mics were integrated into small, self-powered compact flash recording units, and they could withstand wide fluctuations in temperature, humidity, and wind, while still recording huge amounts of audio and metadata over long periods of time (fig. 7).

Some current systems are even arrayed so they can capture signals of equal strength from all points within an imaginary hemisphere using just four carefully aligned and calibrated microphones.[6] There are even models that are submersible, for recording aquatic life, and others that detect and record the ultrasonic

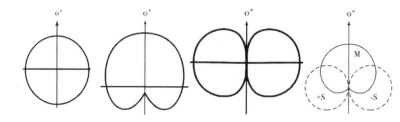

7. Microphone recording patterns: omnidirectional; cardioid; figure 8; mid-side (ms)

signatures produced by bats and insects. Some are so inexpensive and easy to operate that even citizen scientists as young as seven or eight would have no trouble recording in the field.[7]

Digital technologies, with their ever increasing accessibility, value, and ease of use, have given all of us the ability to capture, archive, and analyze increasingly large data sets. They have inspired many of us to more fully appreciate the richness of information contained in natural soundscapes, and also how quickly and profoundly biophonies are being transformed or altogether disappearing. When I began studying soundscapes nearly half a century ago, I could record for roughly fifteen hours and capture one hour of usable material, enough for an album or a sound sculpture installation—a 3-D acoustic work in a public space, such as a museum, that shapes sound in time and space, just as other forms of art like clay or metal sculptures are expressed. Now, because of an intricate and confusing intersection of forces, including global warming, resource extraction, land transformation, anthropophony, air and water pollution, and many other factors, it can take between two hundred and two thousand hours of recording—depending on the condition of the habitat—to capture the same quality material for one usable hour of uncompromised audio.

So where is all this headed? As we've cast new technologies at bio- and geoacoustic domains in order to expose and understand them more thoroughly, we still find ourselves at some distance from fully realizing an informational ceiling of capture and reproduction. For example, there are just too many places left to record along with all their intrinsic environmental permutations. The methods and technologies currently available for analysis are not as precise as they will need to be to understand some of the questions we are beginning to ask. The deeper we dive into this universe, the more questions arise. My guess is that it will be a while before we achieve something like a "perfect" recording system—if we ever can. Such a perfect system would be pocket-sized, water-resistant, better packaged, and could be precisely calibrated. It would contain an interchangeable array of noise-free microphone capsules (similar to camera lenses), with a wide range of selectable patterns, and function under extreme weather conditions for long periods of time; it would embed instant weather, GPS, altitude, and time data, and could be operated remotely in any habitat from any location on the planet.

To date, even though captured sound may be impressive and intriguing, to paraphrase the anthropol-

ogist Gregory Bateson, "The map is not the territory."
Likewise, the recording of a wild soundscape does not
unerringly represent what we think we're hearing, if for
no other reason than our acoustic impressions are often
influenced by the other senses. When we're standing on
a beach listening to the waves, for example, if our eyes
are focused on the breakers offshore, we tend to hear
the distant crashes we're looking at. When we're standing near the water's edge looking at the leading wave of
the water as it rolls up to our feet, we tend to hear the
crackle of bubbles in the surf as the thin membranes
give way and release the gases inside. Even with the
best technologies, those signals and spaces remain only
partially captured and reproduced, while at the same
time providing a vague sense of acoustic and niche
fidelity. Which brings us back again to Walter Murch's
"shadow sense": the ultimate edit and reproduction that
we can only hope comes close to evoking what our ears
hear in its original form at the instant a soundscape was

expressed. Here, too, the future will be framed by a constant striving to achieve a greater degree of realism in acoustic acquisition and reproduction. Since there are no absolute standards by which to judge the result, the goals are still largely subjective and depend entirely on the listener-cum-recordist at one end, and the listener-cum-audience at the other.

One thing is clear: comparable dynamic ranges expressed through the density, diversity, and richness of undamaged habitats are far different from those that have been compromised. Microphones and recorders only capture part of what is present. When I finished a recent lecture, one of the attendees asked how I would express the opposite of biophony or geophony. The answer, sadly, is bioanechoic, a world altogether bereft of non-human sound-producing organisms. Unless, as a world community, we can find some way to mitigate the incredible decline in the world's natural voices, one can only wonder if this is where we are headed.

4 ———————————————

The World Through Soundscapes

With the input of numerous colleagues like Ruth Happel, a primatologist from Harvard University; Stuart Gage, an entomologist from Michigan State University; and the late Ken Norris, a cetologist and conservationist at the University of California, Santa Cruz, I recently came to the important realization that there are an incredible number of disciplines informed by this new branch of bio- and geoacoustics. So far, we have identified many such disciplines, including biology, architecture, medicine, physics, computational mathematics, psychology, literature, resource management, zoology, paleontology, botany, acoustics, natural history, communications, environmental studies, social sciences, theater, business, language, and several more.[1] The following two chapters give predictive voice to those observations.

I have written at length about the ways in which animals gave us our music, powerfully shaping the development of controlled sonic expression, and most likely language as well.[2] Recordings from the field serve as inspirations for new musical literature, as persuasively realized in recently presented works by John Luther Adams and David Monacchi, as well as others. To further test the idea of combining natural soundscapes with orchestral timbres, Richard Blackford, a former composer-in-residence at Balliol College, Oxford, and I composed a symphony that integrates both components seamlessly into a single piece, titled *The Great Animal Orchestra: Symphony for Orchestra and Wild Soundscapes*. The work was premiered in July 2014 by the BBC National Orchestra of Wales, at the Cheltenham Music Festival in the United Kingdom, and recorded for Nimbus Records. The piece sets the stage for future musical structures of this type. But others are clearing paths just as important.

David Monacchi, an Italian composer, naturalist, and music professor, has introduced what he refers to as the *eco-acoustic paradigm*. This phrase begins to explain the innovative and flexible compositional model used for many of his compositions. Monacchi travels to remote sites where he can first immerse himself within the complex mixture of representative soundscapes. He records them faithfully. Then, back in the studio to study the biophonic structure and partitioning that affect nearly all of his recent compositions, he begins to work reverently with the recorded material. To identify these acoustic configurations, he streams highly detailed spectrograms to identify the patterns that connect one signature with another. Then he finds acoustic niches in which to add the elements of his own inspired voice, whether electronic in origin or with haunting melodies he produces on a trans-verse flute. Monacchi enhances his concert performances with the streaming of bio-phonic spectrograms across a large screen (fig. 8). Once he establishes the biophonic niches in the projections, he utilizes them as a musical score and orchestra in which, as a soloist, he weaves his own magical acoustic threads.

The compositions of John Luther Adams, the Alas-kan composer and naturalist, are not easily classified in any traditional sense. To me, the elements that make his work unique lie in the extremes between the atavis-tic and neo-futuristic perspectives of musical chroni-cles. Adams's music, inspired by the wild and majestic spaces and the raw acoustic dynamics that define Alas-ka's landscapes, combines resonant impressions of the constantly changing moods of these mighty settings.

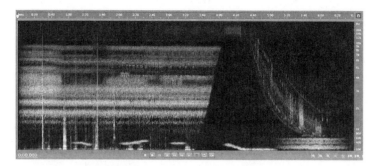

8. Spectrogram of *Integrated Habitats* by David Monacchi

In a small, dedicated room at the University of
Alaska Museum in Fairbanks, Adams created a perma-
nent sound sculpture installation that unfolds along
with slowly changing lighted panels mounted along
one wall with wooden benches opposite. Titled *The
Place Where You Go to Listen,* the installation draws from
a palette of naturally occurring events—geological,
meteorological, and electromagnetic (related to
the earth's magnetic field) among others. All of this
information is fed simultaneously into a device that
transforms the signals into subtle textures of continu-

ously variable electro-acoustic realization. As listeners sit on the wooden benches opposite the wall of subtly changing hues and colors controlled by the soundscape performance, different moods permeate the area. What immediately struck me about this installation, which opened in 2006, was how the combination of the acoustic delivery system and the sounds emanating from the electronic generator greatly expanded the illusion of space, so that when I closed my eyes, I had the impression of being in a much larger environment than the confines of the tiny installation.

More recently, in the spring of 2013, Adams premiered a piece, *Become Ocean,* which was performed by the Seattle Symphony Orchestra. In this work, he incorporated his impressions of climate change—the melting of the tundra, the loss of spring and summer sea ice especially in the Beaufort Sea, the shifting of bird populations farther north with each passing year, and erosion that has been cutting away at coastal Native

American villages throughout many parts of his Alaskan home state. The piece prompted Alex Ross, a reviewer for the *New Yorker* magazine, to remark, perhaps with a bit of hyperbolic irony, "It may be the loveliest apocalypse in musical history."[3] This composition, more tonally centered and clearly structured than the Fairbanks installation, draws on a kind of palindrome structure, where the movements develop themes that build to a climax, then reverse course simply to end where they began, similar to how events might occur under stable conditions in the natural world—life begins with an organized expressive structure, reaches its apogee, then returns to its organic genesis. Remarkably, in this piece Adams has captured a complex overlapping awareness of timing that constantly demands the attention of one's senses to register. For example, not all themes are the same length, range, or texture.

One of Adams's other works was inspired by the idea of the *inuksuit,* man-made stone marker objects

9. Inuksuit, Foxe Peninsula, Baffin Island, Canada; photograph by Ansgar Walk (Creative Commons)

that can be found spread across the Arctic from Yup'ik territory in western Alaska to Greenland (fig. 9).

Much of the tundra is composed of flat open spaces where there are few natural objects to distinguish one location from another, and Arctic peoples erect these inuksuit as landmarks. To record his composition *Inuksuit* for Cantaloupe Music in 2009, Adams installed a thirty-four-piece percussion section in the Vermont woods. Frankly, since the compositional concept was Alaskan-inspired, I would have loved to hear a recording of the performance with natural soundscapes representing acoustic events taking place along the banks of the Kuskokwim River in Alaska's Yup'ik territory, or from the areas surrounding the Askinuk Mountains in the Yukon Delta, or one of the many great and diverse sites in the Arctic National Wildlife Refuge—biophonies and geophonies of the Arctic or sub-Arctic rather than those of the northeastern United States, which to me presents a kind of acoustic disconnect, even though the recording site choice might be otherwise justified as "artistic license" and far more practical.

Natural soundscapes are also either used as inspiration for or included later in music by Antonio Vivaldi, Wolfgang Amadeus Mozart, Claude Debussy, Olivier Messiaen, George Crumb, Alan Hovhaness, Paul Winter, Hildegard Westerkamp, Barry Truax, and some of the programs being produced through the Center for Computer Research in Music and Acoustics (CCRMA) at Stanford University. The natural soundscapes or signature animal recordings (wolves, selected birds, and whales, for instance) in many of these examples were more or less incidental and casual. They are not supplementary, however, in R. Murray Schafer's *Princess of the Stars,* one of the twelve operas from his Patria Series, where Schafer permits the biophony to play a much more direct and integral role in performances.

Far removed from ceremonial halls of culture, Schafer wrote *The Princess of the Stars* to be performed by and on a lake several hundred miles north of Toronto, a venue that requires some planning and effort to travel to for vocalists, instrumentalists, production staff, and audience alike. The preferred natural theater is in northern Ontario, not far from the southern border of Algonquin Park. Instrumentalists from the Toronto Symphony scattered out of the audience's sight throughout the forest surrounding the lakeshore, while the singers, on illuminated boats, begin their musical narratives a little after four on late summer mornings along with the slowly changing sky just before daybreak—emerging from small bays that form the contours of the lake. With the audience seated along the shore, the performers are induced into action by the natural soundscape, led in part by the dawn choruses of birdsong. The implication here is that there is much less need for the appearance of human control over the performance's structure. Although the performance has a prepared compositional protocol, a good deal of the actual recital is left to biophonic chance occurrences at the site.

Schafer has also written an a cappella choral piece about wind, arguably one of the most difficult aspects of the soundscape to convey through musical art. His composition, *Once on a Windy Night,* demonstrates how critical listening can inspire us to incorporate even the most intricate soundscape details, just as Schafer has gracefully captured the essence of this invisible phenomenon with such stunning impact. His books, articles, and compositions have set a high bar, indeed, for Western musical expression related to the natural world. For it is precisely in these types of lyrical and complex connections that future composers are likely to find their muse.[4]

Earlier I described some of the ways in which the contours of the landscape influence the acoustics of a given habitat and thus the individual and collective enactments of the vocal creatures within them. We borrowed some clues from the acoustic properties of these wild spaces and incorporated a few of those spatial concepts into the architectural structures we designed, experimenting with some for their reverberation, others for their hushed, non-echoic qualities, and still others for their ability to block certain types of sound altogether. Church planning from the eleventh through the fifteenth centuries, for example, was rationally conceived, in part, with thick stone walls designed not only to mitigate the street clatter outside, but also to shut out the sounds of the natural world, which in some quarters were considered evil and in conflict with the divinely inspired standards of the human music produced inside. As R. Murray Schafer observed, "Western music is also conceived out of silence. For two thousand years it has been maturing behind walls."[5] For reasons far different from the rationale behind church design of the High Middle Ages, modern architecture still separates unwanted sound from outside. But now the anthropophonic din from within is more controlled—calculated to eliminate superfluous, distracting, and non-informative external man-made acoustic signatures, while at the same time incorporating industry-vetted assurances of more relaxing and less stressful interior ambiences, thus guaranteeing more productivity.[6]

In the realm of future debates over resources, soundscape tools are an important part of the decision-making process. By demonstrating even anecdotally, for example, that the noise created by snowmobiles in Yellowstone National Park had an impact on wildlife and humans alike, we were able to convince certain otherwise skeptical U.S. lawmakers of the need to quiet things down in the wintertime.

The problem reached a head when a loosely knit co-alition of libertarian, anti-environment, and outdoor recreational industry front groups, like the Wise Use movement, promoted expanded land use, unfettered access, drilling, mining, and even private property expansion onto public lands through the 1990s and well into the current century. Specifically and short term, they were advocating for unlimited and uncontrolled snowmobile use in the park.[7]

We did not have time to do a full-blown peer-reviewed study. But if it could be shown, for instance, that the levels of noise affect the experience of a place in the same way that the wrong music and effects of a sound track to a scene in a film distracts from the content and context, we thought it might be possible to create positive conditions for change.

Using a calibrated sound pressure level (SPL) meter set up along the main road between West Yellowstone and Madison Junction, about a mile or so from the West Yellowstone entrance, I recorded the levels of clusters of snowmobiles as they passed by our monitoring stations. A colleague and I measured both two- and four-cycle units and found, in general, that the two-cycle engines were noisier in close proximity to the SPL meters, while the four-cycle engines weren't quite as loud in the near field but generated low-frequency signatures that penetrated far deeper into the landscape—sometimes as much as two to three miles (approximately three to five kilometers), where we had snowshoed into the back country to set up a temporary observation station. With the two-cycle clusters of four or five units each, we measured an average of 91 dBA at fifty feet (fifteen meters) from the center of the road with six different passing groups. (A decibel, or dB, is a measurable unit of amplitude or loudness. The letter A, when used in combination with dB, refers to a weighted average of how the human ear detects audio levels across the entire audio spectrum.) Although we were not able to obtain accurate speed measurements, in several instances they appeared to be exceeding the posted limit of thirty-five miles per hour (56 kilometers per hour). A group of U.S. Congressmen had come on a junket to evaluate the snowmobile noise and pollution issue—ambience otherwise so emblematic of the natural soundscape at that time of year.

Back in a park lodge conference room, I set up a playback system, calibrating the levels so that they measured 91 dBA at the center of the room—the same average level that we had measured from the side of the road. In my PowerPoint presentation, I projected a single photo of a lone wolf feeding on an elk carcass, with lots of ravens scattered about, in a typical snowy winter scene. In my one-minute introduction, I said only that I was going to play two soundscapes of two and a half minutes each, and the audience members were to respond to the projected slide as if they were adding the correct sound to the picture, and that I was going to play the recordings at the same levels at which I recorded them. One example was an anthropophony-free winter soundscape recorded just after daybreak, before snowmobiles were allowed in the park. And the second was a snowmobile cluster drawn at random from those we had recorded over a period of a few days. When I played the first natural soundscape example, everyone in attendance looked calm and their heads nodded affirmatively. But when I played the second audio clip, at the moment the snowmobile soundscape was at its peak of 91 dBA, faces got red, and there were angry shouts to turn down the volume. I didn't. At the end of the demonstration, I said, "This is what you're voting on. From your reactions, you each understand what's at stake and what choice to make."

The National Park Service, after a hard-fought battle with several of the advocacy groups and feedback from visitors untethered to noisy technologies, many of whom had come to the park for the stunning winter quiet it had to offer, finally reached what it thought would be a viable compromise. Instead of noisy and polluting snowmobiles with two-cycle engines, it would allow four-cycle-powered machines guided in limited numbers, down from more than a thousand vehicles per day to the mid-400s around 2005, and more recently into the mid-300s daily.

When taking into account the specter of an entire biome, including the acoustic properties of its landscape, we get a much more complete sense of the living environment. By recording it with calibrated transducer microphone systems designed to capture reproducible signals that transmit not only traditional two-dimensional perspectives, but a full 360-degree replication of horizontal, vertical, and spatial depth, stretching as far as the canopy above, we experience a much more complete sense of a natural community's significant connective details, as in the case of a recording from Corkscrew Swamp in Naples, Florida.

This is not a frequency, amplitude, time-based lens, but a hologram-like spatial view from the perspective of a comprehensive analytic system capable of accurately reproducing the aspects of spread (width), depth, and height based on the relative strength, location, and source(s) of a biome's acoustic signals. Bioacoustic mapping models like these, along with more in-depth post-capture analyses, give us more insight into the acoustic structure of the habitat and the ways in which it operates organically as an expressive unit. Biophonies also define the limits of a habitat's dynamic equilibrium, that is, its dynamic range over time, giving us the agency to tease additional meaning from the signals. For example, with the recorded data in hand, certain types of spectral and temporal cause and effect analysis can show the impact of resource extraction, climate change, land transformation, anthropophony, and weather-related dynamics. All of these, more completely fleshed out, endow resource management teams with more precise forecasts before the sometimes irreversible result of certain activities. A good application paradigm for this process might be a bioacoustic environmental impact review of the route planned for the Keystone XL pipeline, so there is adequate baseline data from which to compare biome status before and after.

5 ———————————————

The Future of the Soundscape

Since the study of soundscape ecology embraces so many disciplines, many questions arise as to how these approaches might be prioritized. How have research communities considered the impacts of soundscapes on whole ecosystems, for example? How will they consider these impacts in the future? What information is revealed by each soundscape component individually and collectively? What are the limits of examining single-species abstractions when removed from the context of the biophony? What new evaluation models might researchers develop to help implement this branch of inquiry? Finally, how does natural sound from any source add to our understanding of our place in the living world? I believe the most intriguing opportunities for study are the applications of biophonies when testing for potential analgesics in human and animal health. In addition, the roles biophonies play in mathematics, natural history, environmental studies, resource management, sociology, climate change, archiving, and the origins of human music and language must be taken into account.

In the sphere of environmental studies, few programs have, until very recently, given much consideration to wild soundscapes. However, it is axiomatic that these topics cannot expect to have a complete operational sense of ecosystems without the necessary attention paid to their collective voices. As I mentioned in the example given in the introduction, our current culture highlights our sense of sight. "Seeing is believing," or so we're told. Or consider the expression "I *see* what you're saying." Conversely, has anyone ever told you that they "hear what you're seeing?"

Seth Horowitz, a neuroscientist at Brown University, suggests in his recent book, *The Universal Sense,* that there is a bit more to the ear than meets the eye. After showing that music and spoken language are unique to humans, he draws on overwhelming evidence to illustrate how important correlated sound is to the Others as well. Then, through a descriptive series of stories, he shows just how much vital information there is for us to absorb and consider to the extent we truly learn how to listen in discriminating ways.[1] Over the past couple of years, as the field has matured, soundscape ecology has become more resolutely interdisciplinary. It assumes that natural soundscapes consist of a combination of both geophonies and biophonies and that they are greatly influenced by a landscape's biological and geological features. It is meant to provide instant feedback as to how well humans are relating to their living environment through the multiple ways in which vocal organisms collectively express themselves. And most important, its proto-narrative informs an astonishingly broad range of subjects.

In the field of bioacoustics, for example, and using the field model I introduced in my 1998 book, *Into a Wild Sanctuary*, the term *biophony* embraces the idea of natural soundscapes more collectively. This grew out of my anecdotal observations going back more than a decade.[2] In parallel, but independently of the U.S. National Park Service collaboration between Stuart Gage and me, Almo Farina and his colleague Nadia Pieretti, at the University of Urbino in Italy, and Jérôme Sueur at the Muséum national d'histoire naturelle in Paris, France, introduced their own variants of the concept in the early 2000s. These were among the first scientists to recognize the value of incorporating entire natural soundscapes as vital diagnostic tools to, among other applications, calculate degrees of habitat viability and noise vis-à-vis certain avian vocalizations. It is no coincidence that many of those studies used new mapping techniques to help visualize their theses.

Though more familiar maps include information pertaining to geographical and land-cover features, often created using technologies such as Geographic Information Systems (GIS) and Global Positioning Systems (GPS), cartographers have yet to fully explore maps of the soundscape, through Bioacoustic Mapping Systems (BMS). In 2007, we experimented by embedding soundscapes into Google Earth maps, providing access for soundscapes for one hundred threatened sites. It was our mission to give voice to the visual aspects of mapping. Google initially agreed, but needed to provide more bandwidth in order to incorporate the stereo audio samples. Because at the time we were not able to provide quality stereo soundscapes at the levels we imagined, we abandoned the project after only a few months, despite well over a million visitors within the first week of its launch. This model for soundscape layering for all types of mapping sites will be one important focus of this rapidly expanding discipline—one we fully expect to be realized in the future.

With the application of mapping technologies like GIS that allow us to visualize and interpret relationships and patterns of a landscape, and LIDAR (technically a fusion of the words *light* and *radar*), a remote sensing application capable of calculating precise distances of landscape contours and defining objects by measuring reflections generated by radar illumination, we are now able to achieve levels of accuracy in habitat evaluation unheard of before 1990.

And new technologies, like the Carnegie Landsat Analysis System Lite, may even have the capacity to penetrate dense forest canopies and habitats not otherwise easily accessible or measurable. The three systems combined will then be able to give us an approximation of physical habitat detail in order to enhance the raw bio- and geophonic data. When united with future holographic soundscape data like that described at the end of the previous chapter, not only will we be able to explore the ways in which biophonics can be used to judge habitat fitness through density, diversity, and richness comparisons, but we will have the capacity to confirm, as nearly as possible, the true value of acoustic mapping just now in its nascent stages. Because bioacoustic mapping defines territories in ways radically different from our typical rational boundaries, this emerging perspective is essential to a clearer understanding of the sonic margins of specific biomes.

Analyzing vast amounts of acoustic data associated with cartographic data will also require more mathematical research. Mathematics, when applied to soundscapes, assesses the current analytical models used to interpret acoustic phenomena in heterogeneous landscapes.

The measured analysis of soundscapes provides

the information necessary to study the relationships between the spatial and temporal distribution
of various species and the environmental characteristics of their habitats, including amounts and
configurations of land-cover types. Both local-scale
spatial variables (such as distance to edge habitats,
or ecotones—transitional zones typically featuring a
combination of open space and forest, where there
tends to be greater density and diversity of wildlife)
and landscape-level spatial variables (such as propor-
tions of different land-cover types in areas surrounding habitat patches) may influence our comprehension of the species composition and distribution of a
region. In our 2001 and 2002 National Park Service
study, we used baseline biophonic recordings as an indicator of habitat fitness for the first time, for instance

 at Crescent Meadow and three other sites
in Sequoia National Park.[3] Healthy soundscapes from most terrestrial habitats con-

tain multiple species' vocalizations with many different time, frequency, and spatial distributions. Among the questions we need to ask are: How are these distributions statistically different, and can we process the signals to determine, with a high level of confidence, that they either are or are not different? Are they all different from one another? Are some correlated, as in timing, either short term (meshing together second-by-second), or long term (taking turns vocalizing)?[4] Some of these questions may be addressed by examining them from other angles.

Over the past several years, another form of detailed visualization has emerged in numerous spectrogram software technologies. Combined with current precision editing software tools, we are approaching a time when it will no longer be necessary to record in the field with parabolic transducer systems, separating vocal organisms out of their natural acoustic contexts. By displaying an entire biophonic spectrogram on the screen, the editing of individual species can be done with the software alone, and then compared with the surrounding acoustic data present in the recording to accurately evaluate the relationships between one vocal organism and others, leaving the overall context intact.

Once we develop a better way to assess and model the soundscape quantitatively, we might further explore its effects. One primary application of sound-scapes might be the improvement of human health. The importance of human and animal health with respect to natural soundscapes springs from anec-dotal evidence of groups living more closely linked to the natural world, particularly those who use the biophonies and geophonies of their environments as healing voices—a kind of non-human shamanistic utterance that is at once both calming and restorative. These groups would have long understood and relied on other healing properties of their environment when they became physically ill or stressed by encounters with outside sources. Several human societies, like the Ba'Aka from the Central African Republic, have engaged the rainforest biophonies and geophonies to help mitigate many types of pathologies.[5] Meanwhile, other studies now under way or being developed address how medical professionals might employ natural soundscapes to relieve symptomatic disorders such as asthma, tinnitus, stress, hypertension, autism, ADHD, and allergies, among other infirmities. A Swiss film, *Nel Giardino dei Suoni* [In the Garden of Sound], made in 2010 by Nicola Bellucci, illustrates the healing properties of biophonies. It features the therapist, musician, and field recordist Wolfgang Fasser, who is both blind and hard-of-hearing, who successfully treats children with Down syndrome, autism, cerebral palsy, and other infirmities, using natural soundscapes and ambient sound.[6]

Preliminary studies analyze the effectiveness of music compared with natural soundscapes as a model for reducing stress indicators, such as high blood

pressure, heart rates and breathing rates, and increased glucocorticoid enzyme levels. We have found
that music often has an effect opposite to desired
outcomes, because of its cultural biases or physical reactions it induces, such as seizures. Anecdotal
evidence indicates that certain natural soundscapes—
biophonies, in particular—may be more beneficial,
perhaps because they are culture neutral.[7] In 2009,
in conjunction with Dr. Mark Tramo, currently the
director of the Institute for Music and Brain Science
at UCLA, but then at Harvard, we generated a proposal
titled *Clinical Research Proposal on Natural Sound, Emotion, and the Brain in Medical and Mental Illness.* The
broad aim was to find more effective ways to alleviate
pain, anxiety, and depression in hospital outpatients
and inpatients by researching, developing, and deliv-
ering the health-related benefits of controlled acoustic
environments. The specific aims were to empirically
test the effect of natural sound on neurophysiologi-

cal and psychological indices of anxiety, depression, and pain caused by common medical and psychiatric diseases in hospital outpatients and inpatients and in a cohort of healthy adult volunteers. The project included a multi-center, randomized, controlled study design using the clinical trial method, and the applicability of clinical research with the potential to improve the quality of health care and control costs through careful comparison of alternative treatments.

Shinrin-yoku, a form of health care and healing in Japanese medicine, literally means "taking in the forest atmosphere" or "forest bathing." Developed during the 1980s, this work has become a cornerstone of preventive health care. In Japan and South Korea researchers have established a robust body of scientific literature on the health benefits of spending time under the canopy of a living forest, seminal components of which are the biophonies and geophonies that the mind engages with during the encounter.[8]

One group in Arizona is even investigating how biophonies and geophonies might help to reduce PTSD (post-traumatic stress disorder) in military dogs returning from the field in Afghanistan and Iraq.[9]

The psychological effects of unwanted, incoherent, or chaotic sound (a.k.a. "noise") on the human condition are well known. They include depression, elevated stress levels, insomnia, headaches, physiological issues, and, in some cases, even death. Studies done mostly in the European Union have begun to explore these phenomena. Some of them have shown that noise can affect the learning abilities of young students in schools adjacent to airports and highways. In an article published in *Noise and Health* in 2010, psychologist Maria Klatte and her colleagues at the University of Kaiserslautern in Germany illustrated a direct correlation between noise in children's environment and their respective levels of achievement.[10] The results of that study were later supported by a paper in 2011 titled

Burden of Disease from Environmental Noise, an examination by the World Health Organization, Europe,
of the effects of noise on children between the ages
of seven and nineteen. This 128-page report states,
"Exposure [to noise such as that of auto traffic and
aircraft] during critical periods of learning at school
. . . impair[s] development and [has] a lifelong effect
on educational attainment," sometimes affecting IQ by
between five and ten points.[11]

Although there is considerable evidence of negative
impact from incoherent and loud acoustic signals, the
emotional consequences of natural soundscapes on
urban-living humans may have a huge positive effect
on our sense of well-being. Assuming the necessary
attention paid to the quality of the field recordings and
subsequent playback systems along with acoustically
controlled and comfortable environments for the lis-
tener, they include soundscapes of waves at the ocean
shore, non-dynamic riparian examples, and even

desert or evening biophonies. These and other realms of investigation explore how these areas of discovery might be further advanced, including childhood development of symptomatic ADHD (a problem I've had to endure my entire life, and which I believe has been partially assuaged by immersing myself in biophonic and geophonic ambient environments), autism, or other problematic behaviors.

As with the clinical study proposal described earlier, programs of this kind will test the efficacy of both anxiety and perceived pain reduction where medical professionals might deliver sound as a palliative. There is anecdotal evidence that soundscapes may have calming applications in hospital settings, either by masking typical hospital sounds or by providing general ambience that will make stays in such sterile environments shorter (because healing might be more rapid) and more pleasant, possibly reducing the need for large quantities of pain-relieving medications. Of course, future research must include double-blind studies to test these theses against other acoustic forms, such as certain types of music.

Even though many of the observations have been anecdotal well into the latter half of the twentieth century, where they do exist, they can provide numerous insights into the perception of natural, rural, urban, or wild sound emanating from both humans and other living organisms. With current supporting analytic tools, such as the GIS, GPS, and other satellite monitoring systems available, we now need to expand the multiple ways in which natural sound can be factored into the narrative of the life sciences.

Soundscape ecology is a subcategory of bioacoustics. It is a field that, in time, will likely confirm that every living organism generates an acoustic signature, and that each unique output signal, individually and as part of a collective expression, has inherent meaning. This discipline involves an exegesis of already recorded species, and an evaluation of techniques to capture those signatures not yet noted, ranging from microorganisms to megafauna. In addition, emphasis is given to the study of whole ecosystems that comprise the soundscapes of larger areas of terrestrial and marine sites. Further inquiry includes an ethology of animal acoustic communication and associated behavior: evolution, ontogeny, and development of acoustic behavior, the physiology of sound production, and neurophysiology. For example, I couldn't understand how birds living in temperate or sub-Arctic climates appeared to discriminate between vocalizations of the same or other species. But Jeff Lucas, an ornithologist teaching and doing research at Purdue University, may have a clue. He recently discovered that black-capped chickadees (*Poecile atricapilla*) actually receive signals as we hear them, but the birds internally pitch-shift (change the frequency of) the signatures up or down so that they are sensed as if they were part of a structured

biophony. Through a process yet to be fully understood, Lucas writes, there is a

> refinement in the auditory system [that] likely enables
> animals to extract species-specific information from the
> background soundscape. The results from our lab and
> others' suggest that the processing of sound provided
> by the peripheral auditory system can allow individ-
> uals to extract very specific auditory signals from the
> array of sounds that a given individual is subjected to.
> For example, through an elaborate method of internal
> pitch-shifting, the song of a chickadee is processed
> differently by a chickadee compared to white-crowned
> sparrows, or compared to a white-breasted nuthatch, or
> compared to a human. The result is that what we per-
> ceive as biophony, or perhaps as a veritable din of noise,
> is not necessarily what a chickadee hears.[12]

The further development of useful field protocols, en-
coding, streaming spectrogram formats, information

storage, search, and retrieval, mixing and post-processing systems, learning more about how non-human animals hear and process acoustic signals, are all part of this future scope.

Beyond its immediately apparent application in biology, biophony is an important aspect in other basic kinds of research. The human impact on climate conditions, for example, is settled science. However, an additional tool used to anticipate future changes in the mix of wild organisms of a given biome as a result of human intervention may be found in additional information expressed through the biophony. Where my wife and I live in northern California, the effect is clear: spring arrives twelve to fourteen days earlier than it did when we moved here in 1993. As of this writing, in 2014, we are experiencing the most pro-longed and severe drought in this region in more than twelve centuries. Our otherwise peaceful rural area is in serious danger of literally running out of water. With or without precipitation, at one spot in the Maya-camas Mountains called Sugarloaf State Park, which marks the geographical boundary between the Sonoma and Napa Valleys, the spring cycle and warmer days have occurred progressively earlier with each subse-quent year; vegetation leafs out sooner, the rate and temporal distribution of precipitation have shifted, and the migrating patterns of birds, once repre-sented by a fairly constant density and diversity, have changed. The result has been a dramatic swing in the mix of dominant species. Their combination in our region was pretty constant in the early 1990s, but new species of birds have been appearing with each succes-sive season since the early 2000s, while the expected ones have diminished in number or have disappeared altogether.

I have been systematically recording a bioacous-tic sampling of this habitat annually since the early

1990s, and the change can be easily heard by listening to three recordings made in April 2004, 2009, and 2014, respectively, at the same location and the same day and time, using a calibrated protocol. In the first two segments, in 2004 and 2009, one can hear in the background the nearby stream, running fully charged; but then in 2014 it is almost silent because of the drought. Accompanying the stream is a biophony of birdsong, from dark-eyed juncos (*Junco hyemalis*), golden-crowned and white-crowned sparrows (*Zonotrichia atricapilla* and *Zonotrichia leucophrys*), California towhees (*Pipilo crissilas*), acorn woodpeckers (*Melanerpes formicivorus*), black-headed grosbeaks (*Pheucticus melanocephalus*), American robins (*Turdus migratorius*), Brewer's sparrows (*Spizela breweri*), red-shouldered hawks (*Buteo lineatus*), pileated woodpeckers (*Dryocopus pileatus*), and wild turkeys (*Meleagris gallopavo*). There is a reduction in density in the birdsong from 2004 to 2009, followed by an almost complete lack of density and diversity in 2014, reflecting the full impact of the drought.

In 1981, at a fairly quiet biome a dozen or so miles northeast of Jackson, Wyoming, I found a site called Spread Creek Pond. It is located in the acoustic penumbra of an otherwise very noisy valley that is home to not only Grand Teton National Park but also the only regional airport located inside a national park boundary. It is a facility with more than twenty departures of private jets and single-engine piston aircraft nearly every daylight hour (and several after dark). In addition to the airport, U.S. Highway 89 cuts right up the middle of the valley toward Yellowstone with all of its vehicular and motorcycle traffic. But Spread Creek Pond is one of the few areas in that region relatively sheltered from disturbances of those types. When I first recorded there, I captured warbling vireos (*Vireo gilvus*), yellow warblers (*Dentroica petechia*), white-crowned sparrow (*Zonotrichia leucophrys*), Wilson's warbler (*Cardellina pusilla*), house wren (*Troglo-dytes aedon*), and the dusky flycatcher (*Empidonax oberholseri*).

When I returned eighteen years later, however, spring was occurring an average of two weeks earlier, according to local naturalists, annual precipitation had changed, and the water level of the pond had bottomed out, leaving a muddy depression. The biophony was unrecognizable. By 2009, the avian diversity consisted of hermit thrush (*Catharus guttatus*), Swainson's thrush (*Catharus ustulatus*), brown-headed cowbird (*Molothrus ater*), black-headed grosbeak (*Pheucticus melanocephalus*), yellow-rumped warbler (*Setophaga coronata*), dark-eyed junco (*Junco hyemalis*), chipping sparrow (*Spizella passerina*), and the usual suspect that occurs nearly everywhere in North America, the white-crowned sparrow (*Zonotrichia leucophrys*). It is especially important to note that my colleague Martyn Stewart, who had joined me for a soundscape workshop in the valley that year, caught what might well be an

 example of a regionally extinct species, the Wyoming toad (*Bufo baxteri*), during a dawn chorus field-recording session.

The toad was not visually identified, the vocalization has yet to be confirmed (some thought it might be a Canadian toad [*Anaxyrus hemiophrys*]), and it has not been heard or recorded at that site since, to our knowledge. But these kinds of recordings demonstrate the potential unleashed when the fine detail of biophonies is examined. At the same time, we have noticed that earlier spring seasons, changes in vegetation, and the influx of different species due to migration pattern variations in the Northern Hemisphere signify the impact of rapid climate change in nearly all biomes now. One of our most important tasks is to find ways to manage and interpret such data.

Among the other disciplines to be examined in the light of soundscape ecology, anthropology can help explain the connections between natural soundscapes and the development of human language, music, physical expression (dance), and social interactions—each one essential to an understanding of human societal development. From the aspect of the soundscape, the investigation meets at the crossroads between us, our acoustic environments, and the impact that natural, rural, or urban acoustic signals have had on us over the course of much of human history.

Different socioeconomic groups will experience the soundscapes of their respective habitats and the impacts they have on their lives as essential to an understanding of place in our ever more complex world. Perceptual differences in soundscapes between normal and hard-of-hearing subjects, and methodology designed to engage and inform the public about the creation, protection, or perception of valued local soundscapes, will be effectively framed into different expressions of cultural dialog.

Because soundscapes establish a strong sense of place for humans, an examination of the ways in which different socioeconomic groups experience their acoustic environment is an essential step in delineating the kinds of surroundings we will choose to inhabit. Above all, the impact of noise—unwanted or random—on quality of life will need further study and interpretation, especially compared with those who claim to be energized by it or otherwise assert that they are not afflicted by these intrusions on everyday existence. Perceptual differences in soundscapes between visually and sonically normal or impaired subjects need to be considered. And methodologies designed to engage and inform the public about the creation, protection, or perception of treasured wild soundscapes or tranquil zones will have to be successfully integrated in the larger cultural dialog if society is to cherish and protect our surviving serene environments.

In addition to basic science, a key beneficiary of

soundscape ecology will be fields such as resource management. As the interrelationships of whole ecosystem soundscapes are revealed, the synergy between acoustic components and the ways in which these links can be implemented as additional tools for measuring habitat viability have become powerful assets. In a matter of seconds, a soundscape reveals much more information from many perspectives, from quantifiable data to cultural inspiration. Visual capture implicitly frames a limited frontal perspective of a given spatial context. But soundscapes widen that scope to a full 360-degree hemisphere—completely enveloping us. Based on the data these records show, accurate projections about habitat sustainability can be made concerning the effects of human enterprise, like resource extraction and land transformation.

Sound design is another application—in theater; broadcast media; public spaces like museums, aquaria, and zoos; and on the Web—it has been part of the expressive fabric as long as each system has been in existence. However, the use of natural soundscapes as a vital and meaningful resource in these media, and above all in public spaces, first became possible with the introduction of analog audiotape systems in the late 1940s. When an obscure engineer at Ampex (whose name is now lost to history) discovered that one could continuously loop audiotape by splicing the two ends together, then triggering the loop to play by the push of a button, a new paradigm for delivery of sound was established and soon became utilized by many composers, including Otto Luening, Vladimir Ussachevsky, Karlheinz Stockhausen, Steve Reich, Brian Eno, as well as Paul Beaver and me (Beaver & Krause, *In a Wild Sanctuary*). That technique is still a main audio protocol considered by exhibit designers to this day. Of course, audio is no longer stored on analog tape. Instead, the sources are now collected on non-degrading silicon chips, the performance of which anyone can initiate.

Despite the fact that dramatic new technologies have been available for almost two decades, many designers of public spaces still opt for the archaic and uninformative push-a-button-hear-a-sound models that first made an exhibit design appearance in 1951 at the California Academy of Sciences, in San Francisco, a few dozen miles from the now defunct Ampex headquarters. Meanwhile, newer delivery systems offer a much wider range of options, many mirroring how natural soundscapes reveal themselves in the wild. They include systems that control levels automatically, identify animal voices in the context of the programs in real time, are interactive with the visitors, and are non-repetitious, meaning that, while content and context remain constant, the programs never recur the same way twice—just as events unfold in the wild. Although these systems have been readily available and are generally more cost-effective than the older push-a-button systems, most public spaces in North America have yet to use these media improvements to any large extent.

The interiors of a good number of public spaces are not designed for acoustic presentations. Natural soundscapes have developed in biomes where species' subtle voices can be expressed without interference and where their voices won't be masked. So if we choose to install them in noisy, reverberant museums, where there is an abundance of air-conditioning, ventilation, and other extraneous noise—or in outdoor exhibits in zoos, where there is interference from aircraft and auto traffic—the idea of a habitat's natural tranquility and successful communication potential is effectively undermined. On the other hand, a future where building designers take into account the theatrical features of the entire space they are creating—addressing not only spatial volume and light, but also acoustic performances—the experiential circle that embraces the natural world will have been completed, assuming the general public begins to value the content and contexts of these exhibitions.

With my growing archive, I often dreamed of integrating these audio files into public space exhibit design as more than just background or "elevator music." I imagined, designed, and subsequently introduced the delivery system described above to be much more engaging and informative. As part of that process, we wanted to better understand how visitors to museums experience sound installations. For the most part, the designers insisted on the push-a-button or audio loop types played through substandard speaker systems. They are easy to install, inexpensive, and, to the designer's mind, deliver some kind of audio component. We were curious to understand what the visitors experienced with these older models.

To find out, we used an exhibit of California habitats installed in one of the rooms of the Academy of Sciences as a study environment. Along one wall of the exhibit, not more than fifteen feet (approximately five meters) from the exit, were five graphics, each representing a different Californian bird. Beside every picture was a button to push, resulting in a short audio clip that represented the bird's vocalization emanating from a small speaker, competing with all the ambient noise in the museum. Just outside the exit portal, we had set up a table with a CD player and earphones, watching and carefully noting only the visitors who pushed all of the five buttons. When those button-pushers got to the exit, we asked them if they would indulge us by taking a short survey. For those men, women, and young adults who responded, we played back all of the five sounds they had just heard a few seconds earlier and asked them to identify them. Of the 125 respondents we engaged over the course of two days, one, an ornithologist, was able to identify all five sounds. Two knew the names of three of the examples. Two were able to identify one each. And 120 were not able to recall a single bird vocalization they had just heard. When we asked the visitors what prompted them to push the buttons, in a multiple choice selection ranging from "I wanted to learn what the voice sounded like" to "I was curious what would happen if I pushed the button," over eighty of them said that it gave them something to do and delivered a result that they were able to control. Here was an indication of instant Pavlovian gratification; it was the least boring and most alluring element in an otherwise static exhibit. Perhaps that was a sufficient result, but most visitors learned nothing from that type of experience.

Sound in constructed spaces is not a new concern. With the exception of some theaters and concert halls, most architectural spaces of the twentieth and early twenty-first centuries have been developed from the perspective of eye appeal. These Ayn Randish triumphs of mind over imagination—which serve as a monument to the architects and the boards of directors whose competitive egos demand them—obviate a visceral or practical understanding of how the acoustic environment in interior spaces will play out. To a large extent, our psychoacoustic reactions to what we experience within those spaces is an afterthought. Curiously, this holds true even in some venues like modern concert halls. The design branches of soundscape ecology will, in the future, be charged with examining more closely the acoustic history of architecture and will also address successful or unsuccessful spaces in terms of their acoustic properties. Special emphasis will be placed on non-theatrical public spaces such as museums, aquaria, zoos, spas, hospitals, and commercial lobbies because of their stated missions and the manner in which architectural engineering has addressed acoustic criteria over the course of recent history. Sound design ideas for the new "greener" LEED (Leadership in Energy and Environmental Design) construction will need to be addressed with site-specific approaches to promote better and healthier urban structures and green-space environments. HVAC, energy conservation techniques, quality-of-life designs, and issues related to enhancing the landscape by bioacoustic mitigation are also under development.

Princeton historian Emily Thompson's book *The Soundscape of Modernity* is a compelling introduction to acoustic design concepts for interior spaces. The narrative examines the changing and challenging history of interior soundscapes during the first third of the twentieth century in America, with an ear to taming and controlling the soundscape of spaces like Boston's Symphony Hall and the sound studios found in Rockefeller Center in New York.[13] With media performances through the 1980s limited primarily to live theater, cinema, and radio and television programming, these issues were manageable, provided sufficient financial resources were available to effect mitigation. However, I believe with the Internet as the most active public

space ever, we need not expend those resources—
financial, material, or human—on the design, construction, and maintenance of more natural history
museums, aquaria, or zoos. To see how the world of all
other living organisms thrives, we only need to install
live-streaming audio and video monitoring sites in the
remaining wild places and project those events directly
onto our media devices. That way we eliminate our
need to maintain captive animals (and be witness to
their resulting pathological behavior) and can instead
show how events unfurl in the real world in real time.
From these online installations, we will be able to
explore the flora and fauna of a site in ways that fit our
levels of interest and curiosity. Low-impact, highly effective models of that sort have yet to be implemented.
Finally, they will spare magnificent wild animals like
killer whales (*Orcinus orca*) from being captured for
Russian, U.S., European, and Latin American theme
parks, separated from their wild pods, and transported

to places like a remote aquarium for the distraction and entertainment of visitors to the winter Olympic Games, another means of asserting our dominance over the natural world.

Outside of architecture, soundscapes have a clear place in art. Western literature has meticulously described the visual aspects of our world. A place can look perfectly healthy to the human eye, but sound as if it is struggling to express itself, like a person trying to speak with a bad case of laryngitis. Trust your ears. Environmental essayist Edward Abbey describes in intimate visual detail the geographical and natural features of the American Southwest, his totem habitat. Isak Dinesen, Beryl Markham, Albert Camus, J. M. Coetzee, and Stephen Biko animated the geographical and political verities of the African landscape. But only a few authors, like John Muir, Aldo Leopold, Willa Cather, Richard Louv, and the Pulitzer Prize–winner Wallace Stegner in his essay series *A Sense of Place,* touch upon the natural soundscapes of their native habitats in any detail. Since the end of the nineteenth century, we have been repeatedly warned about a future absent a deep link to the wild natural and the necessity to preserve what is left. While teaching at Stanford University in Palo Alto, California, Stegner cautioned about a future overrun with human enterprise when he observed:

Something will have gone out of us as a people if we ever let the remaining wilderness be destroyed; if we permit the last virgin forests to be turned into comic books and plastic cigarette cases; if we drive the few remaining members of the wild species into zoos or to extinction; if we pollute the last clear air and dirty the last clean streams and push our paved roads through the last of the silence, so that never again will Americans be free in their own country from the noise, the exhausts, the stinks of human and automotive waste. And so that never again can we have the chance to see ourselves single, separate, vertical and individual in the world, part of the environment of trees and rocks and soil, brother to the other animals, part of the natural world and competent to belong in it. . . .

. . . We simply need that wild country available to us, even if we never do more than drive to its edge and look in. For it can be a means of reassuring ourselves of our sanity as creatures, a part of the geography of hope.[14]

As the recording technologies improved and I and my colleagues began to amass larger and larger quantities of data, many of us found it necessary to document what we were accumulating. The key to maintaining and successfully developing these collections is archiving —the storage and retrieval of the recorded material and metadata. Generally, natural sound collections fall into two categories: one is attended capture, and the other consists of unattended, or automated, field recording. Attended recordings—which I mostly prefer because it gets me out into the wild for extended periods and makes me feel good—is a model predicated on the recording of one or two sites at one time and for relatively short continuous periods like, for instance, a dawn or evening chorus typically lasting one and a half or two hours at a stretch. The recordist is usually never too far from the recorder and microphones he or she has set up. The potential downside of attended recording is the possibility of affecting the behavior of some avian or mammalian species by one's presence or missing events that occur at times when we are not recording. The upside is that the volume of data and metadata from that paradigm is usually manageable. That said, improvement of storage for retrieval and analysis is definitely a subject for ongoing discussion and development.

On the other hand, with the introduction of durable, inexpensive storage, and self-powering field recording technologies that also note weather, time, and location, it is now possible to install a number of temporary units in a given habitat, set them to record examples for specific times and durations, and leave them on site to do their work for weeks at a stretch before returning to mine the data (which can also be accessed and streamed by satellite). Taking pride in my work, I once announced at a science meeting that I had collected more than forty-five hundred hours of material over forty years. During a break, and not to be undone, the convener of the conference took me aside and announced, "Whadya mean?" with his jaw muscles tensed but overjoyed at the opportunity to gain a competitive edge. "A couple of months ago I set up eight devices each at four different sites and collected nearly sixteen thousand hours of audio, in just four weeks!" The downside of that mass of data is pretty clear. How does one deal with it?

Scientists at Michigan State University's Remote Environmental Assessment Laboratory (REAL) may have an answer. Under the leadership of Stuart Gage, the REAL team has devised a software program that automates the process of cataloging acoustic sensor observations into the REAL digital library.[15] The library facilitates access and analysis of collected acoustic sensor observations from multiple stations. It then provides a report on current library status and the mechanisms that enable the selection, extraction, and analysis of acoustic data to support investigations on automating species census as well as measuring diversity and disturbance. The numeric and symbolic search mechanisms that drive the system combine with unsupervised learning techniques to ease the problems inherent in the retrieval of acoustic information, including recordings and processed data, all of which relate to and aid the researchers' goals.[16]

The information contained in natural soundscapes gives us an ontological understanding of the biological niches we inhabit, yielding numerous prisms through which to view our relationships to the non-human critter world. For some the association is spiritual and cultural. For others, the link is merely a numbers game—scientific, abstract, and secular. With sound-scape ecology, both masters are served.

Because soundscapes are becoming recognized as a special means of engagement with the natural world, in 2001 the National Park Service incorporated them as a resource with the same level of importance as, say, wildlife and the parks' flora. It was to be a new lure for visitors. The good news is that visitors collecting soundscapes in the parks and then factoring them into post-production product is one of those endeavors that essentially does no harm; no trees need to fall, no landscape needs to be transformed, ground does not need to be broken to obtain liquid, gas, or metals, and relatively little, if any, energy is used. It is truly a lovely, gentle way to mesh one's life with that of all other living organisms.

Currently used in film, public spaces (theaters,

museums, zoos, aquaria, and so on), and Web-based and other media delivery systems, each soundscape has inherent value as an entity to be studied, developed, and factored into a wide range of performance pieces. Recent trends suggest that archival library holdings and academic curricula will be expanded into a diverse range of disciplines. Special emphases on the pairing of research, science, technology, and the arts in multiple media offer unlimited opportunities for expression.

When we experience the natural soundscape of a given habitat, whether desert, riparian, tropical, or temperate woodland, or the seashore, we are always faced with the question of its intrinsic significance weighted with ethical and moral questions all subsequently tied to an expression of and reverence for life.

A line by the novelist Ellen Glasgow—"Preserve, within a wild sanctuary, an inaccessible valley of reveries"—informed the title of one of my record albums, a book, and the name of our small organization.[17] Compared with the enormous quantity of visual metaphors and imagery, relatively little has been written about aural characteristics. I suspect this is largely because there aren't many descriptive ways to portray sound (again, even here I am using a graphic word, *portray,* to describe an acoustic event). To close this gap, we need to reexamine acoustic references from pre-biblical times to the present, allusions where sound is portrayed as part of the spectacle of human experience. The inclusion or dismissal of acoustic references often illustrates the degree to which we succeed in balancing the sentient experiences we describe through our various means of communication.

Regarding the elemental impact of soundscape ecology in general and natural soundscapes specifically, my colleague, naturalist, and close friend Ruth Happel once remarked: "They helped shape music, and if we lose the sounds of the wild, then we will also lose an important inspiration and resource for the arts. When you hear a chimp drumming in the woods against a buttress, that is the origin of drums. When you hear the melody of a bird, that is the origin of our own melodies. If they are gone, our own music will wither."[18]

At this moment in history there are essentially two distinct outcomes to the work being done in the field of soundscape ecology. The first portends a somewhat ominous landscape void of living sounds. Absent our recognition of and willingness to stem our incessant consumption of resource-intensive products and the progressively devastating effects of the fossil-fueled energy necessary to produce it, we will soon enough find ourselves living in smaller land-based areas (due to sea level rise), with all the nasty competitive human activities required to survive, and few, if any, truly wild areas left to explore—scenarios well extrapolated by a diverse range of authors. For example, the University of California professor Ozzie Zehner (author of *Green*

Illusions), the Canadian novelist Margaret Atwood, and the environmental activist and author Bill McKibben all point out from different perspectives the numerous large and small tragedies that become imminent absent our willingness to overcome our compulsive drives to consume. Pope Francis, in his November 2013 exhortation on the state of the world economy, acknowledged: "In this system, which tends to devour everything that stands in the way of increased profits, whatever is fragile, like the environment, is defenseless before the interests of a deified market, which becomes the only rule."

The second possible outcome is a bit more optimistic. The field of soundscape ecology is expanding faster than we can keep pace and the amount of useful data being collected is staggering. As Happel noted, "Arrays of recorders, and the ability to automate data analysis, will provide measurable trends in soundscape dimensions. So it is imperative to identify the sources of funding to acquire these tools. Funding is generally biased toward the 'hard sciences,' so the more soundscape ecology researchers are able to provide consistent ways to compare data by different researchers, the better."[19]

Those drawn to this discipline will need to address several looming problems. The first is to find and record as much bioacoustic data in the few remaining pristine habitats as possible—those locations with minimal anthropophony or evidence of human enterprise. The second issue is to reverse the paucity of trained human resources necessary to adequately capture and analyze soundscape material by encouraging more individuals worldwide to engage with this endeavor. The third is to gain a clear understanding of the information being transmitted by assessing the complex interrelationships that exist between vocal organisms in a given habitat. Last, there is a need for finely tuned and calibrated field technologies and agreed-upon protocols for capturing, analyzing, and managing data.

Even though life expresses itself in many ways, it is extremely difficult to apply meaning to the vast store of information contained in collective bioacoustic utterances, largely because we're engaging with the subject so late in the game and there's very little left to explore that is truly wild. Nevertheless, I cannot emphasize enough the need to capture audio snapshots of these remaining sites wherever they happen to be. Inexpensive and easy-to-use recording technologies help make possible the exploration of natural soundscapes by anyone who can operate an iPad, smart phone, or PlayStation. It's a thrilling realm of discovery for citizen-scientists as young as five years old or for any non-hearing-impaired octogenarian. As a result, new findings, such as the specific characteristics that define the unique relationships between individual species' vocalizations within a given habitat, are being made with ever increasing frequency. Activities like these can only add to the dedicated research of soundscape ecology groups at institutions such as Michigan State University (Envirosonics Department), the University of Urbino (Italy), the Muséum national d'histoire naturelle (Paris), and East Anglia in the United Kingdom. Online communities provide vast troves of valuable information from the perspective of many skill levels, and include everything from the latest field technologies and methodologies to special sites for recording, archiving, applications of the recorded material into performance pieces, and science leads.[20]

As they continue to be deciphered, these motifs will not only reveal exciting new layers of the earth's historical, scientific, and cultural past, but also provide new comprehension in the varying degrees of our human impact on wildlife and wild habitats so we can plot more productive paths in the future. Most important, they convey a prescient sense of what lies ahead because of how these changes can now be envisioned, quantified, and extrapolated. What will the associations between our endeavors and their direct impact on wildlife be? And how will that impact be expressed bioacoustically?

The wisdom of many of the best naturalist writers and thinkers has postulated that humans are unlikely to comprehend all of the complex signals offered to us through the ever-shifting sands of the wild; we are not designed to decipher its entire bag of enigmas. Perhaps, though, as we begin to unravel a few omens expressed through biophonies and geophonies, we will benefit from a worthy glimpse into that captivating universe, one more complete than what now exists.

Natural soundscapes are a declaration of the present and they entreat us to think and respond likewise. Nevertheless, we seem bound to manipulate time by attempting to capture impressions of the moment and then project aspects of them beyond their primal impact into notions before or after. Part of the human transaction is to notice ways of showing deference to a mutable thriving world on its own terms; not how we rationally find ways to spin the past or imagine moments to come. To the extent that we are able to accomplish that simple task, our lives will unfold with much more wholesome outcomes.

From my experience, it appears as if we find ourselves stuck midway between a past and a future where chance intimate encounters with what remains wild actually underscore our inability to escape the present. Our history confirms that we have the limited capacity to postulate and imagine aspects of the days ahead of us. In order to do that with some degree of confidence, though, we will need to more fully understand what exists. Natural sounds that define the field of soundscape ecology are the voices we need to heed closely. For they are balanced somewhere between creation and destruction—and we silence them at our own peril.

Which will win out? My fondest wish would be to survive long enough to hear the wonders that will be discovered through the multihued prism of this fabulous medium and to have faith that wild soundscapes will still be heard by and inform others of our species, young and old, just as I have come to know and love them.

Notes

INTRODUCTION

1. The term *shadow sense* was coined in 2012 by Walter Murch, an Academy Award–winning sound designer and audio historian.

2. Almo Farina, *Soundscape Ecology: Principles, Patterns, Methods, and Applications* (Springer, 2014 edition), 6.

3. Bernie Krause and Stuart Gage, *SEKI Natural Soundscape Vital Signs Pilot Program Report: Testing Biophony as an Indicator of Habitat Fitness and Dynamics,* National Park Service (3 February 2003), 2.

4. Bernie Krause, *Into a Wild Sanctuary* (Berkeley, Calif.: Heyday Books, 1998).

5. Krause and Gage, *SEKI Natural Soundscape Vital Signs,* 2. The term used originally was *anthrophony,* meant to specify human-generated sound as a component of the soundscape. It is a word that we have been using, incorrectly, since the early 2000s. Only recently, while giving talks in France during the summer of 2014, was it pointed out to me by our French hosts in Quimper (Pierre Mens-Pégail) and Paris (Jérôme Sueur) that the Greek prefix *anthro* meant cave, and not human, as we mistakenly assumed. Since we were not addressing the sounds that caves produce, in order to correct the term we needed to add a "po" and modify the spelling to *anthropophony*.

6. Jérôme Sueur, "Cicada Acoustic Communication: Potential Sound Partitioning in a Multispecies Community from Mexico (Hemiptera: Cicadomorpha: Cicadidae)," *Biological Journal of the Linnean Society* 75, no. 3 (March 2002): 379–394.

7. Bernie Krause, "Anatomy of the Soundscape: Evolving Perspectives," *Journal of the Audio Engineering Society* 56, no. 1/2 (January–February 2008).

CHAPTER 1. THE BIRTH OF THE SOUNDSCAPE

1. Carl Troll, "The Geographic Landscape and Its Investigation," in *Foundation Papers in Landscape Ecology*, ed. John A. Wiens, Michael R. Moss, Monica G. Turner, and David J. Mladenoff (New York: Columbia University Press, 2007), 71–101; originally published as "Die geographische Landschaft und ihre Erforschung," *Studium Generale* 3, no. 4/5 (1950): 163–181.

2. R. Murray Schafer, *Tuning of the World* (McClelland & Stewart, 1977).

3. Christopher Cokinos, *Hope Is a Thing with Feathers: A Personal Chronicle of Vanished Birds* (Tarcher, 2009).

4. Bernie Krause, *The Great Animal Orchestra: Finding the Origins of Music in the World's Wild Places* (Little, Brown, 2012).

5. Bernie Krause, "The Niche Hypothesis: A Virtual Symphony of Animal Sounds, the Origins of Musical Expression, and the Health of Habitats," *Soundscape Newsletter,* World Forum for Acoustic Ecology, 6 June 1993.

6. Emily J. Mockford, Rupert C. Marshall, and Torben Dabelsteen, "Degradation of Rural and Urban Great Tit Song: Testing Transmission Efficiency," *PLoS ONE* 6, no. 12 (2011), e28242 DOI: 10.1371/journal.pone.0028242; Nadia Pieretti and Almo Farina, "Application of a Recently Introduced Index for Acoustic Complexity to an Avian Soundscape with Traffic Noise," *The Journal of the Acoustical Society of America* 134, no. 1 (July 2013): 891–900, DOI: 10.1121/1.4807812.

7. Marla M. Holt, Dawn P. Noren, and Candice K. Emmons, "An Investigation of Sound Use and Behavior in a Killer Whale (*Orcinus orca*) Population to Inform Passive Acoustic Monitoring Studies," *Marine Mammal Science* 29, no. 2 (2013), first published online 12 September 2012, DOI: 10.1111/j.1748-7692.2012.00599.x.

8. Louis Sarno, *Bayaka: The Extraordinary Music of the Babenzélé Pygmies* (Ellipsis Arts, 1996).

9. Shierry Weber Nicholsen, *The Love of Nature and the End* of the World: The Unspoken Dimensions of Environmental Concern (MIT Press, 2003).

CHAPTER 2. THE CHALLENGES

1. Paul Shepard, *The Others: How Animals Made Us Human* (Island Press, 1996).

2. Louis Sarno, *Bayaka: The Extraordinary Music of the Babenzélé Pygmies* (Ellipsis Arts, 1996).

3. Christopher Small, *Musicking* (Hanover, N.H.: Wesleyan University Press, 1998).

4. Don Young and Richard Pombo, letter written to Gale Norton, Secretary of the Interior, dated 21 November 2003.

5. Bernie Krause, *Wild Soundscapes in the National Parks: An Educational Program Guide to Listening and Recording,* National Park Service, 9 February 2002.

6. Charles Jurasz, personal correspondence, June 1994. Also, Bernie Krause, *The Great Animal Orchestra: Finding the Origins of Music in the World's Wild Places* (Little, Brown/Hachette, 2012), 191.

7. Sheela McLean, NOAA Fisheries News Release, Alaska Regional Office, 14 May 2007.

8. Charles M. Jurasz and V. P. Palmer, "Distribution and Characteristic Responses of Humpback Whales (*Megaptera novaeangliae*) in Glacier Bay National Monument, Alaska, 1973–1979," National Park Service report, 1981; Christine M. Gabriele and Tracy E. Hart, "Population Characteristics of Humpback Whales in Glacier Bay and Adjacent Waters, 2000," National Park Service report.

9. Scott Creel et al., "Snowmobile Activity and Glucocorticoid Stress Responses in Wild Wolves and Elk," *Conservation Biology* 16, no. 3 (June 2002): 809–814.

10. Suzanne Goldenberg, "Climate Change Is Good for You, Says Ultra-Conservative Heartland Institute," *The Guardian,* 9 April 2014 (www.theguardian.com/environment/2014/apr/09/climate-change-report-heartland-institute-debunk-ipcc); Todd Wilkinson, *Science Under Siege: The Politician's War on Nature and Truth* (Johnson Books, 1998).

11. Richard Louv, *Last Child in the Woods: Saving Our Children from Nature-Deficit Disorder* (Algonquin Books, April 2008).

CHAPTER 3. TECHNOLOGICAL PROGRESS

1. John Burton, "Ludwig Koch: Master of Nature's Music," *Wildlife Sound,* Autumn 1974, reprinted on Wildlife Sound Recording Society (www.wildlife-sound.org/journal/archive/koch.html).

2. Bernie Krause, *The Great Animal Orchestra: Finding the Origins of Music in the World's Wild Places* (Little, Brown, 2012).

3. Bernie Krause, *Wild Soundscapes: Discovering the Voice of the Natural World* (Wilderness Press, 2002).

4. Ron Streicher and F. Alton Everest, *The New Stereo Sound-*

book, 2nd edition (Los Angeles: Audio Engineering Associates, 1998).

5. Duke Ellington, *Stereo Reflections of Ellington*, CD released December 1993 on the Hall of Sermon label.

6. Soundfield and AmbiSonic arrays.

7. The Song Meter SM2+ field recorder and other models described above have been designed and manufactured by Wildlife Acoustics, Concord, Massachusetts.

CHAPTER 4. THE WORLD THROUGH SOUNDSCAPES

1. Bernie Krause, "Anatomy of the Soundscape: Evolving Perspectives," *Journal of the Audio Engineering Society* 56, no. 1/2 (January–February 2008).

2. Bernie Krause, *The Great Animal Orchestra: Finding the Origins of Music in the World's Wild Places* (Little, Brown/Hachette, 2012).

3. Alex Ross, "Water Music: John Luther Adams's 'Become Ocean,' at the Seattle Symphony," *New Yorker,* 8 & 15 July 2013, 92–93.

4. Krause, *The Great Animal Orchestra*.

5. R. Murray Schafer, *Voices of Tyranny, Temples of Silence* (Arcana Editions, 1993).

6. Emily Thompson, *Soundscapes of Modernity: Architectural Acoustics and the Culture of Listening in America, 1900–1933* (MIT Press, 2004).

7. Sharon Beder, "The Changing Face of Conservation: Commodification, Privatisation, and the Free Market," in *Gaining Ground: In Pursuit of Ecological Sustainability*, ed. David M. Lavigne (Guelph, Ontario: International Fund for Animal Welfare, and University of Limerick, Ireland, 2006), 83–97.

CHAPTER 5. THE FUTURE OF THE SOUNDSCAPE

1. Seth Horowitz, *The Universal Sense: How Hearing Shapes the Mind* (Bloomsbury, 2013).

2. Bernie Krause, "Bioacoustics: Habitat Ambience and Ecological Balance," *Whole Earth Review*, no. 57 (Winter 1987).

3. Bernie Krause, Stuart H. Gage, and Wooyeong Joo, "Measuring and Interpreting the Temporal Variability in the Soundscape at Four Places in Sequoia National Park," *Landscape Ecology* 26 (August 2011), DOI 10.1007/s10980-011-9639-6.

4. Nick Miller, of Harris Miller Miller & Hanson, Inc., Burlington, Mass., personal correspondence, October 2013.

5. Louis Sarno, *Bayaka: The Extraordinary Music of the Babenzélé Pygmies* (Ellipsis Arts, 1996).

6. Nicola Bellucci, *Nel Giardino dei Suoni*, a film produced by the Soap Factory, Basel, Switzerland, 2010.

7. Oliver Sacks, *Musicophilia: Tales of Music and the Brain* (Vintage, 2008); Bernie Krause, *The Great Animal Orchestra: Finding the Origins of Music in the World's Wild Places* (Little, Brown/ Hachette, 2012).

8. Yuko Tsunetsugu, Bum-Jin Park, and Yoshifumi Miyazaki, "Trends in Research Related to 'Shinrin-yoku' (Taking in the Forest Atmosphere or Forest Bathing) in Japan," *Environmental Health and Preventive Medicine* 15, no. 1 (January 2010): 27–37.

9. IDD Paw Jaws, Inc., Julia Jaworski, president (www.idd-PawJaws.com).

10. Maria Klatte, Thomas Lachmann, and Markus Meis, "Effects of Noise and Reverberation on Speech Perception and Listening Comprehension of Children and Adults in a Classroom-like Setting," *Noise and Health* 12, no. 49 (2010).

11. Lin Fritschi et al., *Burden of Disease from Environmental Noise: Quantification of Healthy Life Years Lost in Europe,* World Health Organization publication, March 2011.

12. Jeff Lucas, personal correspondence, via email, 6 October 2009.

13. Emily Thompson, *The Soundscape of Modernity* (MIT Press, 2004).

14. Wallace Stegner's "Wilderness Letter," originally written to the Outdoor Recreation Resources Review Commission in 1960, and subsequently printed in *The Sound of Mountain Water* (first published 1969; reprint, Penguin, 1997); quoted in "Eco-Speak," *Writing for Real: Writing in the Service-Learning Contact Zone,* Stanford University (http://web.stanford.edu/~cbross/Ecospeak/wildernessletter.html).

15. The recorded data and metadata can be accessed through the Web site of the Remote Environmental Assessment Laboratory, Michigan State University (http://lib.real.msu.edu).

16. Eric P. Kasten et al., "The Remote Environmental Assessment Laboratory's Acoustic Library: An Archive for Studying Soundscape Ecology," *Ecological Informatics* 12 (2012): 50–67.

17. Ellen Glasgow, *The Sheltered Life* (Charlottesville: University Press of Virginia, 1994; originally published 1932).

18. Ruth Happel, personal correspondence, 24 December 2013.

19. Ibid.

20. Naturerecordists@yahoogroups.com, worldlistening @yahoogroups.com.

Further Exploration

TO READ

Farina, Almo, *Soundscape Ecology: Principles, Patterns, Methods, and Applications,* Springer, 2014

Hendy, David, *Noise: A Human History of Sound and Listening,* HarperCollins, 2014

Keizer, Garrett, *The Unwanted Sound of Everything We Want: A Book About Noise,* PublicAffairs, 2010

Krause, Bernie, *The Great Animal Orchestra: Finding the Origins of Music in the World's Wild Places,* Little, Brown/Hachette, 2012

——, *Wild Soundscapes: Discovering the Voice of the Natural World,* Wilderness Press, 2002

——, "Anatomy of the Soundscape: Evolving Perspectives," *Journal of the Audio Engineering Society* 56, no. 1/2, January–February 2008

Schafer, R. Murray, *Tuning of the World,* McClelland & Stewart, 1977

——, *A Sound Education,* Arcana Editions, 1992

——, *HearSing,* Arcana Editions, 2005

——, *Voices of Tyranny, Temples of Silence,* Arcana Editions, 1993

Shepard, Paul, *The Others: How Animals Made Us Human,* Island Press, 1996

Sider, Larry, ed., *Soundscape: The School of Sound Lectures,* Wallflower Press, 1998–2001

Turner, Jack, *The Abstract Wild,* University of Arizona Press, 1996

TO WATCH

Nel Giardino dei Suoni [The Garden of Sound], a film by Nicola Bellucci, Cineworx/Soap Factory, 2011, with English, French & German subtitles

TO LISTEN

Over 50 soundscape titles from a variety of field sound recordists: http://wildstore.wildsanctuary.com

Chris Watson's BBC series: *A Life in Sound* http://www.bbc.co.uk/programmes/po1qcldf

Blackford, Richard, and Bernie Krause, *The Great Animal Orchestra: Symphony for Orchestra and Wild Soundscapes.* BBC National Orchestra of Wales, Martyn Brabbins, Conductor. Nimbus Records, 2014. (The first full symphonic composition to integrate natural soundscapes as a component of orchestration.)

TO VISIT

Rather than give specifics, imagine a place to visit where there is little or no human noise and lots of wildlife still present. You'll need to do a bit of soul searching and research to find those spots. Be sure to pick a time of year and location that meets those criteria. Pack a decent audio recorder, strap on your backpack, buy a ticket and go there. There aren't many places left that are completely wild (as in no roads, no signs, no well-groomed paths, and nothing to buy). One of my favorite places to visit and record is Alaska. I've been there maybe fifteen times and have rarely gone to exactly the same place more than once. It's a large area with very few people. Just what I love the most . . . getting lost for weeks at a time without a GPS, iPhone, or anything else to remind me of the daily distractions I usually encounter.

TO JOIN

British Library of Wildlife Sounds (www.bl.uk/reshelp/find helprestype/sound/wildsounds/wildlife.html)

Macaulay Library (Cornell University) (www.birds.cornell.edu/page.aspx?pid=1676)

Michigan State University Envirosonics program (http://www.cevl.msu.edu/envirosonics)

Department of Basic Sciences and Foundations, The University of Urbino

Campus Scientifico "Enrico Mattei," 61029 Urbino, Italy (http://ecoacoustics.sciencesconf.org/)

naturerecording@yahoogroups.com
naturerecordists@yahoogroups.com

Wild Sanctuary (www.wildsanctuary.com)
World Forum for Acoustic Ecology (http://wfae.proscenia.net/)
World Listening Project (worldlistening@yahoogroups.com)

Index